D1565766

# Terrorism, Radicalism, and Populism in Agriculture

# TERRORISM, RADICALISM, AND POPULISM IN AGRICULTURE

LUTHER TWEETEN

**Luther Tweeten** is Professor Emeritus of Agricultural Policy and Trade at Ohio State University. He is editor of four books and author or co-author of seven books and over 500 journal articles and published papers. He is a former President and a current Fellow of the American Agricultural Economics Association. Recent awards include the Charles Black Award from CAST, the Henry A. Wallace Distinguished Alumni Award from the Iowa State university College of Agriculture, the Distinguished Scholar Award from Ohio State University, and the Lifetime Achievement Award from the Southern Agricultural Economics Association.



This text was produced from camera-ready copy provided by the author.

Iowa State Press
2121 State Avenue, Ames, Iowa 50014

Orders: 1-800-862-6657
Office: 1-515-292-0140
Fax: 1-515-292-3348
Web site: www.iowastatepress.com



Printed on acid-free paper in the United States of America

First edition, 2003

Library of Congress Cataloging-in-Publication Data

Tweeten, Luther G.
Terrorism, radicalism, and populism in agriculture / Luther Tweeten.—
1st ed.
p. cm.
Includes bibliographical references and index.
ISBN 0-8138-2158-4 (alk. paper)
1. Agriculture and state—United States—Citizen participation.. 2. Populism—United States. 3. Environmentalism—United States. 4. Agriculture—United States—Costs. 5. Environmental policy—Economic aspects—United States. 6. Agriculture—Environmental aspects—United States. 7. Agricultural innovations—United States—Public opinion. 8. Radicalism—United States. I. Title.
HD1761 .T864 2003
338.1—dc21

2002014176

The last digit is the print number: 9 8 7 6 5 4 3 2 1

# Contents

# Preface

After the terrorist attack on New York's World Trade Center on September 11, 2001, I began writing this volume. While that attack was by foreign terrorists, terrorism is not foreign to the United States and to agriculture in particular. This book is mostly not about potential dangers posed to agriculture and food by foreign terrorists. It is about the ubiquitous populist and radical groups and individuals of domestic origin who have imposed massive costs on agriculture and the nation.

Although the "911" attack was my motivation for writing this book, the root motivation goes back four decades. During that time it was my good fortune to interact quite often with agricultural populists and radical activists. I have had numerous personal discussions and debates and have written extensively about activists' issues. I joke that I am frequently asked by activists to joust because I so often lose. In reality, agricultural populism and radicalism are no cause for levity.

Agricultural populists and radicals include antiglobalists, autarkists, radical environmentalists, and animal welfarists. These individuals and groups take few human lives, but they are not benign. They promote a lot of what I call the "soft terrorism" of false information and ideologies. A few individuals misled by such misinformation unfortunately go on to the next step–engaging in vilification and violence toward the object of their enmity. Fortunately, violence is rare and it is mostly directed at property. Still, terrorist attacks in agriculture perpetrated by Americans are numerous and damaging.

Radicalism and populism in agriculture are often inserted into politics, bringing bad public decisions that cost Americans billions of dollars in lost income each year. My subject is those costs and the false ideologies and myths that caused them. In addition, I fear the massive losses.in truth, income, and lives that would be imposed on rich and poor people alike around the globe if proposed populist and radical nostrums were implemented.

I am speaking here to all students of agriculture, whether they be laypersons, faculty, classroom students, agribusiness persons, politicians, or civil servants. I believe that we as a nation invite difficulties when we depart from our Enlightenment philosophy rooted in reason and free inquiry. This book recognizes that, while science and benefit-cost ratios are highly useful tools in making decisions, they alone are inadequate and must be supplemented by values that decision makers bring to the table. Values deserve intense scrutiny, and reliance on gut feelings–emotions–to make decisions is a formula for tyranny.

I am grateful for the help of many people in preparing this volume. Shawn Carpenter helped me to cope with the word processor; colleagues Barry Goodwin, Douglas Graham, Ian Sheldon, Douglas Southgate, and Carl Zulauf critiqued various chapters; and my wife, Eloyce, offered patience and encouragement. John Tomasek, a great friend of agriculture, helped identify sources of information. Of course, none of these persons is responsible for shortcomings of the book.

# 1

# Introduction to Radical, Populist, and Terrorist Agriculture

As I write this, on September 12, 2001, it is one day after the terrorist destruction of the twin towers of the World Trade Center in New York City. Thousands of people have died, and I am reminded of the human capacity for evil. This terrorist act was perpetrated by members of al-Qaeda based far away in the mountains of Afghanistan. But al-Qaeda is hardly the sole proprietor of terrorism. Unfortunately, it is an activity as American as, well, agriculture and apple pie. Agriculture? Isn't agriculture supposed to be about peaceful family farms, sleepy rural communities, and law-abiding citizens?

Terrorism in this book means unlawful acts of violence against property or people designed to accomplish political objectives through fear and intimidation. Fortunately, agricultural terrorism has not killed many people. This volume is mostly about the "soft terrorism" perpetrated by alternative agriculture radicals and by populists who peddle "snake oil" (misinformation, myth), hate, and property destruction in agriculture. The progression is usually in that order. Only a few who are fed the snake oil of erroneous information and false ideology turn to hate or moral outrage. And fewer still of those who hate turn to violence.

My belief is that confronting misinformation can combat some of the unfortunate thinking and behavior that follows. I also presume that post-September 11 Americans will be less tolerant of misinformation and terrorism promulgated by or toward agriculture than in the past. The acts of agricultural radicals and populists have been frequent and of long duration. The consequent massive costs to Americans are detailed in this book.

Because the subject is sensitive, I want to make clear what this book is not about. I have only praise for pathbreaking "alternative agriculture" innovators such as the Spray brothers of Mt. Vernon, Ohio; Richard Thompson of Boone, Iowa; and the late Robert Rodale of Emmaus, Pennsylvania. They seek to reduce synthetic chemical inputs and protect the environment. It is not that I always agree with their conclusions. Rather, having known them personally, I am impressed with their commitment to science, public scrutiny of their work, and good humor and goodwill toward those who hold different views about how best to produce food. Similarly, I regard highly much of the work of the Wallace Institute for Agricultural and Environmental Policy, the Leopold Center, and similar institutions using science to find different and sometimes radically new and better ways to produce food and serve the environment.

I have steadfastly called for reform of agricultural policy, for safe food and water, for care of the environment, for innovation, for change, and for appreciation of the virtues of people on farms. I have no quarrel with peaceful protest.

So why is this book largely a condemnation of radical agriculturalists (RAs) and populist agriculturalists (PAs)? The book is about *radical* alternative agriculture advocates, where radical is defined as groups and individuals who promote their vision for agriculture with misinformation, deceit, fear, extortion, hate, and/or violence. Their discourse is characterized by reckless and willful disregard for facts, logic, analysis, property, and people.

Agricultural populists tend to be less extreme and indeed are mostly close to the agricultural establishment (see chapters 7 to 9). But in pursuit of their self-interest (higher prices and income for farmers), PAs too assault facts and logic. Economists only rarely use the term "evil," and I will follow that tradition, but I will argue that radicals and populists engage in behavior that causes individuals and society to make unwise decisions. This volume will show that the cost to society of such activity from RAs and PAs is huge in destruction of trust, national income, property, and lives.

Actions of the September 11 terrorists and of Adolph Hitler, Joseph Stalin, Chairman Mao, Ted Kaczynski, and Timothy McVeigh were not written in the stars or in their genes. Their actions were the product of nurture, not nature. They were nurtured on a diet of misinformation and an ideology of distrust and hate that resulted in violence.

I and other Americans prize our freedom of expression, dissent, and peaceful protest. That freedom extends to dissenters. But those who peddle "snake oil," hate, and violence deserve to be challenged in the marketplace of ideas and need to face the force of the law when they break it. This book focuses mainly on the former, that is, the marketplace of ideas. In challenging misinformation and false ideology perpetrated by today's radical agriculturalists (RAs) and populist agriculturalists (PAs), the premise is that a more informed public will make better decisions and fewer individuals and groups will "progress" into the various stages of anger, hate, and violence.

## CONFRONTING CONTENTIOUS ISSUES

Numerous contentious issues must be confronted on these pages. Some extremists contend that preserving the environment and small family farms are such worthy ends that they justify nefarious means to change the dominant culture. On this issue, I like what the Apostle Paul says in his epistle to the Romans (3:8) "Why not say–as we are being slanderously reported as saying and as some claim that we say– 'let us do evil that good may result?' Their condemnation is deserved." Part of the difficulty is that the ends pursued by radicals are not worthy. But even pursuit of worthy ends does not justify slaughter of truth or lives.

Others contend that attacking those who peddle snake oil, hate, and violence is sinking to their level. I do not accept that contention. I accept Edmund Burke's (here paraphrased) advice that evil will flourish where informed people do nothing to challenge misinformation. Nonetheless, issues of what is misinformation and, indeed, what is right and wrong must be and are addressed in this volume.

## WHY ABANDON REASON NOW?

Reason seems to be abandoned in contemporary tribalism, religious fundamentalism, postmodernism, populism, and in animal rights and environmental terrorism. Is not that a refutation of the dominant Enlightenment philosophy on which this nation, if not modern Western civilization, was founded? Shouldn't modern science, education, and information technology make the world more rather than less rational and civil?

Part of the answer to these rhetorical questions is that America and other parts of the world are not necessarily becoming more irrational and ideological. Modern communication technology such as the Internet only makes today's few radicals and populists more visible and effective.

Part of the answer also is that education does not necessarily create tolerant, thoughtful people. Anyone who has spoken on controversial topics on campuses as I have knows that students–even at prestigious colleges that you would expect to be bastions of academic freedom and free speech–readily deny free speech by shouting down speakers who do not conform to their preconceived ideas.

Some other possible explanations, elaborated in subsequent paragraphs and chapters, for lapses of civility and reason include:

- The search for simple and certain answers to complex questions that inevitably arise in a fast-paced, dynamic, complex world. For many, technology creates a pace of change threatening their values and their jobs. They would like to return to a simpler, more predictable existence that perhaps never was. They would like to return to the immutable answers provided by religious creeds, secular ideologies, and agrarian (if not hunter-gatherer) societies.
- Boredom. The world suffers from a vacuum in respectable alternative ideologies that promise intellectual fulfillment and a better life. That is, once-promising "isms"–socialism and Marxism–that provided the alternatives to capitalism have proven to be dead ends. People see global problems of poverty, illiteracy, hunger, AIDS, injustice, and corruption. They are impatient for solutions. The lack of a viable alternative ideology to democratic-capitalism leaves a void that many fill with nihilism. An academic community is especially prone to unrest in such an environment.
- The limitations of science. Many of the important questions of life cannot be answered by science. In other cases, science has been looked to for answers but has provided wrong answers due to inadequate data, incompetent analysis, or other reasons.

Paradoxically, Americans view agriculture as unique–as the ideal of bucolic tranquility in a troubled world. This volume will disabuse people of that notion. Americans are known for being moralistic, and it is tempting for pilgrims who find no moral bedrock elsewhere to find it in agriculture. Thus, the counterculture revolution of recent decades has not eluded agriculture. The issue is how the revolution is apparent in propaganda, ethical philosophy, and behavioral patterns regarding agriculture.

## OUR DIFFERENT WORLD

Radical activism and terrorism are especially effective means for small numbers of people to achieve objectives that they could not hope to achieve in the market or voting booth in an advanced democratic-capitalist nation. Modern industrial and postindustrial economies are notable for their intricately interconnected transportation and communication networks. An interruption of transportation or food supply at critical points can cause much trauma. These characteristics coupled with the large scale of firms and specialization of consumer and producer activities mean that just a few individuals can wreak havoc and hence extort concessions from society far out of proportion to their numbers and resources.

Today's complex emerging technologies, products, and institutions seem inscrutable to the masses and are easily demonized by media-savvy activists. The Internet facilitates coordination of protest or terrorist activity. The relatively small cost of activism and terrorism to perpetrators and the large costs they can impose on victims and society create an asymmetric threat. That threat is perceived to be an attractive private benefit-cost ratio for the unscrupulous. Thus, we can expect to see more of this darker side of human activity in the nation and in agriculture.

On a recent visit to Monticello, Virginia, I was taken by the guide's comment that Thomas Jefferson required three days to cover the 100 miles from his home to Washington, D.C. Where can one go today in three days? The answer is "Anywhere in the world!" The world is "smaller" today than was northern Virginia at the time of Jefferson, and it is much smaller for communication than for travel. So each of us is touched by what happens abroad as well as at home. Globalization is an inescapable reality. The tide rolls in however loudly the modern King Knutes rail against it. It follows that nihilistic ideologies and terrorist acts can be coordinated and spread quickly around the world today with modern transportation and communication to touch anyone and everyone.

Another important development is the principle growing out of the 1960s and 1970s that "agriculture is too important to be left to agriculturists." Student protestors in high moral drag, after condemning the Vietnam War, inevitably turned their indignation to other points of concern, among which were agriculture. Protestors concluded that food and agricultural policy in the United States and indeed in the world needed reform and revolution.

It is irony that out of these and other origins of globalization tens of thousands of protestors have been able to gather at various international summits in Seattle, Washington; Gothenburg, Sweden; Prague, Czech Republic; Davos, Switzerland;

Washington, D.C.; Quebec, Canada; Genoa, Italy; and other venues since 1999 to rail, sometimes violently, against globalization! The protests were mostly at meetings of the International Monetary Fund, World Bank, World Trade Organization, and World Economic Summit. Many of the protestor's concerns were about agriculture. Key issues were trade, genetic engineering of foods, third world debt and food security, and the international economic system–especially the role of multinational corporations and profit in that system.

RAs frequently address concerns that they feel are not and cannot be properly addressed by markets. They suffer no lack of emotional commitment to what they sincerely consider to be causes of overwhelming importance to the future of humankind. In fact, it is this deep emotional commitment that frequently blinds them to facts and logic in their headlong urge to do "good." Markets do not serve them; they must find recourse in the political system. Frustrated by their lack of voting power and finances, and impatient with a seemingly tone-deaf public that does not listen to their concerns, they distort analysis and turn to emotion to demonize the targets of their vehemence.

A key belief of many RAs is embodied in the statement from the Organization for Competitive Markets, or OCM (March 2000, p. 1), that "emotion is an absolute necessity for reason." Many persons would agree that values are integral to virtually any analysis; the mischief begins when that emotion or value is hate.

Consider a few inflammatory quotes from the *Newsletter* of the OCM:

- These [agribusiness] mergers are anticapitalistic and profascism (March 1999, p. 1).
- [The] power of corporate agribusiness in the marketplace…is the true cause of family farm devastation (July 1999, p. 1).
- [Agribusiness] industry structure becomes what it is today–a steamroller destroying independent agriculture (January 2000, p. 1).
- The Perverted Capitalism [of corporate agribusiness] will fall only when awakened citizenry gains critical mass to overwhelm the crony-protectors (June 2000, p. 1).

OCM did not exist in the early 1980s, but numerous other groups spewed plenty of emotional rhetoric demonizing agribusiness. For example, Doug Jenness, in *The Militant*, a publication out of Dallas, Texas, stated in 1980 that "[Small farms are failing] not because they are 'inefficient' or 'too small'. Rather it is because they are not paid for a big portion of the value they create; it is stolen from them by the owners of the banks, land, and trusts."

By creating a climate making agribusiness people less than human, such scapegoating pushes some people over the edge to violence. Bruce Brown (1989) wrote a poignant actual account of a basically decent but stressed Iowa farmer who killed his banker, wife, neighbor, and finally himself. The farmer blamed the banker for his son's ill-timed plunge into the Iowa land market in the early 1980s. Farmers killed bankers in several other unrelated cases around the nation in the 1980s.

As I watched the horror of the terrorist attack on the World Trade Center unfold, I was reminded of the attack by eco-terrorists that could have taken the life of my friend and fellow agricultural economist, Professor David Schweikhardt. He was working in his office on New Years Eve in 1999, just one floor below the biotechnology office firebombed that evening by the Earth Liberation Front (ELF). The ELF apparently was not impressed by the fact that the biotechnology office was engaged in an effort to help poor people in third-world countries raise more food. Fortunately, Professor Schweikhardt escaped unharmed. ELF's next attack may kill. Shortly before I arrived on the University of Wisconsin campus to spend the year as a visiting scholar in 1972, antiwar protesters blew up a physics laboratory in the early morning hours, killing a hapless graduate student who was working some extra hours.

ELF is the most active terrorist group in the United States, claiming credit for acts of violence in 12 states as of July 2001. It claimed credit for destroying the University of Washington's Center for Urban Horticulture, which among other things, contained one-quarter of the world's known population of the endangered flora called showy stickweed. In addition to agricultural experimental plots, ELF has vandalized logging machinery and equipment, Forest Service stations, highways, ski lodges, large homes, and sports utility vehicles. The large spikes pounded into trees by ELF saboteurs "explode" the saws cutting logs into lumber, endangering the lives of sawmill operators.

The hate and conspiracy themes permeating radical and populist agriculturalists often distort their reading of information from scientific sources. For example, in its *Newsletter*, OCM sought to justify the need for emotion in reasoning by egregiously misinterpreting the research of Antonio Damacio, a neurologist at the University of Iowa. In his book *Descartes' Error,* Damacio (1994) describes the pathological behavior of people made socially dysfunctional by injury to the prefrontal area of their brain from blunt force trauma or stroke. Their cognitive capabilities remain nearly intact, but they lose much more than their emotions. They lose their accumulated learning from the socialization process, including their orientation toward the future. Thus, it is quite incorrect to say that they lose their ability to function in society because they lose their emotions. They lost their common sense.

Damacio was dealing with mental pathology, which OCM ascribes to those individuals who don't agree with its ideology. For example, in its *Newsletter (*March 2000, p. 1*)*, OCM contended that a former economist colleague of mine "and others of his ilk need to realize that their myopic version of 'objectivity' may be closer to pathology." Their only "pathology," however, was failure to support OCM's animus against agribusiness. Thus, a pattern of paranoia characterizing many PAs and RAs begins with frustration over existing circumstances. Unable to muster a reasoned response through existing institutions, they lash out with misinformation, demonizing alleged wrongdoers. In some instances, this process culminates in terrorism as in the case of the Animal Liberation Front, the Earth Liberation Front, and a number of organizations protesting at meetings of the IMF, World Bank, and the World Trade Organization. Thus, radical organizations can become not just prosecutors, judges, and juries, but also executioners.

Free enterprise economies have a proud record. On the whole, capitalism deserves praise rather than enmity. Emotions of overall, generalized love or hate for an economic system are not useful, however. Businesses that perform well deserve praise; those such as ADM that fix prices or such as Enron that deceive investors and workers need to be prosecuted to the extent of the law. Inadequate laws and regulations need to be strengthened.

The private sector will not alone formulate nor enforce a proper regulatory framework in a free enterprise economy. That is an important job for the public sector. When the public sector fails in this critical task, the blame rests as much or more with the public sector than the private sector. Hence, what RAs label "market failure" is in fact often "government failure."

## SOME CHARACTERISTICS OF RAS AND PAS

RAs are by no means homogeneous, but they share a number of characteristics. Although most Americans would place RAs on the left in the political spectrum, RAs have ultraconservative or reactionary dimensions. Many do not like to see a *new* opening of trade, a *new* method of plant breeding, or a *new* laborsaving machine.

RAs are rarely found in and rarely originate from the conventional agricultural establishment comprised of commercial farmers, their organizations, agribusiness (farm input supply and product marketing firms), land-grant colleges of agriculture, agriculture committees in Congress, and the U.S. Department of Agriculture. Postmodern philosophy detailed in chapter 2 provides the intellectual cover and conspiracy theory binding RAs together. University students and well-educated, affluent suburban housewives are especially drawn to RA activity. Student support tends to hale from colleges of arts and humanities and from sociology rather than from colleges of agriculture, engineering, and physical sciences. Commitment to the postmodern philosophy detailed in chapter 2 seems to be inversely correlated with the rigor of science in the various disciplines on college campuses. The income elasticity of demand for RA activity is high; poor people are so absorbed in the vicissitudes of the here and now that they don't have time or money to worry whether their chicken is raised on or cooked on the range.

As noted earlier, RAs are sincere, deeply committed persons who are highly moralistic in their thinking. They are more likely than is the general public to draw their moral beliefs from secular or from liberal religious traditions. With a political philosophy to the left of center, they call for government to severely circumscribe if not end the power of market capitalism and the free enterprise system. As liberals, they fear that no one is in charge to ensure that technology and globalization will be shaped to improve the lot of humans.

RAs and PAs alike are troubled by *anomie*, to use a term from sociology. That is, they lack trust in and feel lack of control over their sociopolitical environment. The *liberal dilemma* is that the big government essential to implement RA's vision for humans has had a dismal record of doing so–witness the disastrous environmental record of former and current centrally planned countries. Thus, RAs have a love-hate relationship with government, and use violence and misinformation to attain their goals partly because they do not trust conventional democratic processes characterized by heavy reliance on the vote and the popular media.

RAs are at their best operating in the error zones of conventional agricultural sciences. Social and many other scientists typically test hypotheses at the 5 percent level; that is, on average they will err by rejecting a correct hypothesis 5 percent of the time. (Social scientists also err by failing to reject a false hypothesis on average perhaps 5 percent of the time in repeated experiments.) Thus, RAs have an opening to be right 5 percent of the time in addressing the same issues as those addressed by conventional science. For example, in their advocacy of laetrile derived from apricot pits as a cure for cancer but found to be ineffective by scientists, RAs stand some small chance of being right. Science cannot *prove* that a particular food containing a genetically modified organism is safe for *all* persons for *all* time.

Fear, overstatement, and deception used by Upton Sinclair in *The Jungle* and by Rachel Carson in *Silent Spring* are considered unethical by most scientists, but aroused the nation to make some worthy reforms. In an open society such as America, reforms come, though perhaps with some delay, without muckraking by the likes of Sinclair and Carson. Nonetheless, humility and caution are warranted before scientists reject the claims of RAs–they are sometimes right and conventional science is sometimes wrong. The odds are heavily on the side of conventional science, however. Those who seek remedies for personal or economic ills in the nostrums prescribed by RAs do so at their own peril, not always because the RA snake oil is harmful but because it diverts attention from more effective remedies.

## RAS AND PAS GET AN ASSIST FROM THE PRESS

RA and PA causes, however unworthy, receive wide exposure and adherents in part because of help from the press. I can only conjecture why television and print reporters often side with the populist and radical agenda for agriculture. Surveys indicate a strong liberal bias among television and print reporters, drawing them to RA and PA positions (e.g., antibusiness) preponderantly left of center. Goldberg (2002, pp. 123, 124) reported results of Freedom Forum and Roper Center polls indicating that 50 percent of Washington, D.C., journalists identify with the Democratic Party and only 4 percent with the Republican Party. Fully 89 percent of the journalists said they voted for Bill Clinton in 1992, more than double the 43 percent of nonjournalists who voted for Bill Clinton.

The Washington, D.C., journalists may not be representative of the national press, but the liberals in the press culture are sufficiently pervasive to view their thinking as mainstream–as so sound that they do not need to be identified as liberal or even as opinion. Thus, nonagricultural journalists too readily accept myths that agriculture is an ever-worsening environmental basket case, that food is unsafe to eat, that small family farms are more economically efficient and environmentally sound than larger corporate farms, that agribusiness rigs markets to keep farmers poor, and that international trade is bad for the environment, consumers, and the economy.

Why do many in the press embrace these myths? Reporters' backgrounds play a part. Perhaps it began with a left-brain versus right-brain orientation. That is, youth with strong verbal and intuitive skills are drawn to arts, letters, and journalism where they are exposed to postmodernism as explained in chapter 2, while youth with strong reasoning and numeric skill were drawn to sciences, engineering, and business. Tendencies might be reinforced because youth in the liberal arts and humanities are taught by professors paid less than those in hard sciences. It is not difficult for those "oppressed" professors to impart to students some of their frustration with the establishment. Or maybe it started with an upbringing in a liberal, "social gospel" church where members are admonished to "go out and do good" by righting injustices, however much misery (ruined property and lives) such activism causes. And finally, the liberal bent may come from nothing more sophisticated than the need for *news* to be dramatic reports of impending economic and environmental doom rather than a prosaic but factual discourse on an ever-improving standard of living and quality of life.

## THE HIGH COSTS OF RADICALS AND POPULISTS

The hate, violence, and misinformation peddled by RAs and populists can be costly. Populist economic policies followed in many developing countries are immiserating millions of people. Millions of additional people mostly in poor countries would become or remain chronically short of food if RAs' advice were heeded to avoid synthetic pesticides and fertilizers, genetically modified organisms, and international trade.

The costs of radicalism and populism are not confined to poor counties. Americans spend over $8 billion annually for organic foods based largely on misinformation that such foods are better for people and the environment. The weight of evidence is that organic foods are worse than conventional foods for the environment and no better than conventional foods for nutrition, taste, and safety. Americans spend $20 billion or more annually on farm policies that serve populist political goals but do not serve goals such as economic equity or efficiency that are consistent with improving the well-being of people.

Agrarian (agricultural) populists often work in concert with RAs and labor unions to form a formidable public relations and political front. In contrast to RAs, agrarian populists arise mostly from agriculture and are often part of the agricultural establishment. Compared to RAs, radical populists have a long history in the United States, going at least back to English settlers in the 1600s, as discussed in chapter 9. Agrarian populist organizations such as the National Farmers Union, National Farmers Organization, and the American Agriculture Movement have origins in protest and minor violence, but over time have evolved to work primarily through conventional political and media channels to achieve their objectives.

RAs and agrarian populists find common cause on a number of issues including trade protection, legislation to help organized labor, and opposition to globalization. Perhaps the single most common object of paranoia pervasive among RAs, agrarian populists, and other counterculture organizations is corporate business, including the international business and banking establishment.

**OUTLINE**

Chapters 2 through 6 deal with RAs. Chapter 2 reviews their philosophic orientation to help us understand why they think the way they do. Subsequent chapters critique the positions of major groups representing the four categories of RAs listed below.

1. *Antiglobalists*. These include persons and groups opposed to trade liberalization (autarkists), to multinational corporations, and to multilateral agencies including the World Trade Organization, World Bank, and International Monetary Fund.
2. *Radical environmentalists*. Persons and groups promoting radical analysis and radical solutions for perceived world problems of overpopulation, excessive energy use, greenhouse gas emissions, and "unsustainable" agriculture.
3. *Luddites*. Persons and groups opposed to economic development and modern technology, including genetically modified organisms for food, and industrialization and mechanization of agriculture.
4. *Animal rightists*. Those who promote radical solutions and violence to provide animals with rights equal to humans' rights.

Chapters 7 through 9 address the origins and philosophy of agricultural populism, critically examining their creeds and positions on economic issues of agriculture. The final chapter (chapter 10) ties together highlights of earlier chapters and examines the potential for agroterrorism from foreign sources.

Treatment of international terrorism such as that of the al-Qaeda organization is brief because, fortunately, international terrorism has neither sprung from nor targeted American agriculture. Nonetheless, vigilance is critical, as emphasized in chapter 10. One lesson of this volume is that snake oil, hate, and violence affecting American agriculture is mostly homegrown.

**REFERENCES**

Brown, B. *Lone Tree.* New York: Crown Publishers, 1989.

Damacio, A. *Descartes' Error*. New York: Grosset/Putman, 1994.

Goldberg, B. *Bias*. Washington, DC: Regnery Publishing, 2002.

Organization for Competitive Markets. *Newsletter*. Lincoln, NE: OCM, 1999 and other issues.

# 2

# Understanding How Radical Agriculturalists Think: Postmodernist Philosophy

Professor Carrol Bottom, an illustrious agricultural extension economist at Purdue University, was asked upon his retirement what he had learned from decades of experience. He replied that he had learned that two intelligent people could look at the same set of facts and reach very different conclusions regarding their meaning. He was impressed with the extent to which our thinking is conditioned by our "priors" or values.

Few people realize the critical contribution of unspoken philosophic systems to their own and to others values and perceived self-interests. Philosophical systems influence how scientists and laypersons alike choose and define problems, how they analyze those problems, how they interpret analysis, and how they resolve those problems.

Someone asked what I have learned from four decades of work as a professional agricultural economist. I replied that I was especially impressed with the ability of pretty much everyone to "rationally" justify his/her self-interests. Self-interests undoubtedly help to explain why radical agriculturalists (RAs) and conventional agriculturalists (CAs) think differently. Self-interests also influence our choice of philosophy to order our thinking. This chapter is about philosophical systems, especially postmodernism.

Western philosophy has followed two major branches or systems in the nineteenth and twentieth centuries. One is the *analytical*, modern, or Anglo-American tradition. The other is *continental* philosophy, out of which postmodernism has developed.

## ANALYTICAL PHILOSOPHY AND THE ENLIGHTENMENT

Notable foundations of analytical philosophy include the logic of Aristotle, the rationalism of René Descartes, the empiricism of Francis Bacon, the pragmatism of John Dewey, the statistical inference of R. A. Fisher, and the positivism of Lionel

Robbins and Auguste Comte. The Anglo-American analytical philosophical tradition is exemplified by the scientific method, weaving facts and logic in a framework of testable hypotheses to produce knowledge that is as reliable as possible–given human frailties. The scientific method confronts testable hypotheses with empirical analysis. Analytical philosophy is a philosophy of optimism–that application of science and technology in an environment of rational thinking will lead to a better, albeit imperfect, world. It continues to be the principal philosophic orientation found in the nation's universities.

Thomas Jefferson and other of the nation's founders were steeped in Enlightenment philosophy, a variant of the analytical school of philosophy. The Enlightenment continues to be the core philosophy of the nation to this day. It had origins in the philosophy of Reneé Descartes emphasizing reason over emotion in making decisions, the philosophy of John Locke emphasizing that the ideal world lies in a natural order of no collective restraints on individual actions, and in the utilitarian philosophy of Adam Smith, emphasizing that humans, by their acquisitive instincts, are led to a better life by the invisible hand of the market.

The Enlightenment called for decisions regarding society to be made on the basis of objective reason and rationality rather than of subjective judgment and emotion. Enlightenment philosophy is inseparable from Logical Positivism, which holds that objective science ideally should base conclusions on verifiable facts and tested hypotheses rather than on value judgments–the latter being subjective and hence unreliable by definition. That is, knowledge must be subject to verification in terms of experience.

Analytical philosophy in its various manifestations has dominated applied economics since at least Adam Smith. More significantly, the analytical philosophical tradition has so dominated U.S. science that many U.S. scientists are unaware of the continental system of western philosophy. Though marginalized for decades, continental philosophy, culminating in postmodernism, has gathered considerable momentum since the 1960s. Its influence is being felt in science, technology, and policy. This chapter reviews the genesis of that philosophy and its contribution to radical movements.

## THE GENESIS OF POSTMODERN PHILOSOPHY

I review the thinking of four "grandfathers" before examining the positions of three "fathers" of postmodern philosophy.

### Grandfathers of Postmodernism: Marx, Freud, Kierkegaard, and Nietzsche

The grandfathers of postmodernism were from continental Europe and made most of their intellectual contributions in the nineteenth century. Karl Marx was a political economist and Sigmund Freud a psychiatrist. Soren Kierkegaard was a Lutheran minister. Fathers of postmodern philosophers examined later in this chapter drew from their thinking. Among grandfathers of postmodernism, the German philosopher Friedrich Nietzsche probably had the greatest impact on the fathers of postmodernism.

*Friedrich Nietzsche* (1844-1900). Nietzsche worked with three propositions. One was *perspectivism*, the view that facts cannot be separated from interpretations. Objectivity is beyond human capability because the mind cannot know truth in an objective sense. Minds are indeed useful, but according to Nietzsche invariably flawed because they cannot separate facts from human error and moral values, which inevitably are subjective. It follows that making sense of science requires a careful look at presuppositions–the priors analysts bring to their work.

If all perspectives are subjective and hence flawed, what perspective is society to follow? Nietzsche's (1973, pp. 33-54) position was that no source of knowledge was authoritative. Sources of knowledge won ascendancy based on which ones were backed by the holders of power. Thus, perceived "truth" depended on power. Real truth, if it existed, was not bestowed by princely or divine power, but was relative and subjective. It depended on circumstances.

The second principle of Nietzsche is *overcoming* or liberation. Recognition that moral values are subjective and that rights can only be interpreted in their social context frees the observer to break from the bondage of false views to see society more clearly if still subjectively.

Nietzsche's first two principles engendered a third principle of *nihilism*, a term he originated. Evidence of a God as creator had been dispelled in Nietzsche's view by Charles Darwin's theory of evolution. *Perspectivism* and *overcoming* likewise freed society from a Creator providing moral imperatives. Nietzsche argued against "the archaic categories of good and evil, long since unmasked as the sentimental traces of power relationships" (quoted from Anderson 1998, p. 65). Primacy of individual moral and methodological authority coupled with rejection of a higher, universal moral authority constituted *relativism*. Absent absolute truth, one moral position could be judged only in relation to other moral positions.

Nietzsche's nihilism has some similarities with Anglo-American analytical philosophy holding that science is merely a set of unrejected hypotheses (Ghebremedhin and Tweeten 1994, p. 12) that can hardly be mistaken for absolute truth. However, while analytical philosophy uses statistical analysis and other means to record and reduce error, Nietzsche's nihilism (denying any objective basis for truth) has sometimes been interpreted to mean that the search for knowledge is fruitless, that there is no meaning or purpose in existence, and that institutions are too corrupt to be worth preserving. An example of the latter is the violent Russian nihilism movement of 1860 to 1917.

*Karl Marx* (1818-1883). Marx's dialectical materialism views progress toward utopian socialism in historical stages. In the final stages, the *thesis* of capitalism confronts the *antithesis* of labor to bring a *synthesis*–a dictatorship of the proletariat culminating in a withering of the state, leaving pure socialism to supply each according to his or her needs and ask from each according to his or her ability. Marx (1970, pp. 352-56) viewed corporate capital as the power controlling society. It aggrandizes profit by exploiting labor. Government, legal, and kinship institutions are dependent on financial support from business, and thus serve the interests of capitalists. In summary, dialectical (historical) materialism is a theory of *society*, despite its economic determinism.

That many postmodern philosophers began as Marxists may seem a paradox because postmodernism rejects "grand narratives" of which Marx's dialectical materialism may be the grandest of all. Many postmodern thinkers rejected Marxism because it was apparent by the 1960s that Marxist economies were unable to transcend the dictatorship phase, and even that phase was not a dictatorship of the proletariat (workers). Obtaining true socialism through communism was seen as another grand illusion. Postmodernists rejected Marx's strict economic determinism, historical dialectical materialism, dictatorship of the proletariat, and capitalist downfall theories. However, many postmodernists retained Marx's positions regarding class struggle, business exploitation, worker alienation, and the conspiracy of corporate business to obtain and use power to control society for profit.

*Sigmund Freud* (1856-1939). Sigmund Freud was the founder of psychoanalysis. He proposed three subsystems of the human psyche: the *id* (primal instincts, drives, wants), the *ego* where decisions are made, and the *superego* that contains the norms internalized from parents and culture (Freud 1960, pp. 9-29). The ego is the "processor" or "traffic cop" of the mind, reconciling the impulses of the id to the constraints of the superego. All manner of neurotic or compulsive-obsessive behavior presumably results from constraints imposed on the id.

Some postmodernists generalized Freud's theory of the mind as a science of society. In extending the psychoanalysis of Freud to sociology, postmodernists rejected Freud's preoccupation with the primacy of sexual motivation (libido). But they retained belief in the role of subconscious, subjective elements in society and science, and in the seemingly neurotic behavior of many otherwise normal people. Thus, the psyche could never be freed from the seemingly irrational and subjective influences of the unconscious (subconscious) id repressed by the conscious superego of norms derived from an oppressive society.

*Soren Kierkegaard (*1813-1855). Kierkegaard was the father of Christian existentialism, a way of thinking that grew out of his efforts as a Lutheran minister in Denmark to reform the state church. His struggle for reform was lonely and subjective. "Truth" for him came from his inner feelings and intellect rather than from the establishment or other traditional sources (Kierkegaard 1962, ch. 3). Armed with that "truth", the individual persisted in bringing change however hopeless seemed the struggle. Jean-Paul Sartre, a French secular existentialist, also held that the great questions of life had to be answered by turning inward; answers could not be found in the great "nothingness" of the outside world against which existence is set.

**Fathers of Postmodernism: Lyotard, Foucault, and Derrida**

Several French philosophers carried the foundation of postmodernist philosophy laid by the above thinkers to fruition in the twentieth century. Numerous individuals could be cited, but three prominent French philosopher fathers of postmodernism were Lyotard, Foucault, and Derrida.

*Jean-Francois Lyotard* (1924-1998). Lyotard (1979, p. xxiv) defined *postmodern* as "incredulity toward meta-narratives." Defunct grand narratives according to Lyotard include Enlightenment progress, Christian redemption, science, and socialism. He viewed scientific knowledge as "a kind of discourse" (p. 3). Interpreting Lyotard, Anderson (1998, p. 25) stated, "the defining trait of the postmodern condition is the loss of credibility of these meta-narratives...science became just one language game among others, it would no longer be viewed as...the progressive unfolding of truth."

Lyotard (1979, pp. 75–82) viewed the undoing of science as self-inflicted—science had come to serve the power of the state and big business (capital). Lyotard, initially a militant on the far left, argued that both a market economy and socialism were bent on accumulating power and capital. They must be rejected not because they were irrational but because they were rational toward inappropriate ends (Anderson 1998, p. 27). Eventually, Lyotard mellowed in his disdain for the market on the grounds that it was a product of economic evolution validated by performance rather than a grand narrative created by humans. "The triumph of capitalism over rival systems, he now argued, was the outcome of a process of natural selection that predated human life itself" (Anderson 1998, p. 33).

Lyotard is not the only postmodern to grudgingly accept capitalism. Postmodern philosophers Deleuze and Guattari (1983, p. 153) reluctantly conclude " it is capitalism that is at the end of history." Unlike Marx's socialism, however, it is not perceived to be utopia. Deleuze and Guattari (pp. 264-67) refer to it as a "schizophrenic" capitalism that commoditizes everything with markets and prices, destroys the extended family, and, through multinational corporations, controls nation states to serve business interests.

*Michel Foucault* (1926-1984). Foucault referred to written and spoken *discourse* metaphorically as the multifaceted dimensions of human activity. An emerging discipline or science gathers information from numerous sources, often informally with no attribution to authors, and from no dominant conceptual framework or authority. "All these constitute a sort of anonymous system" (Foucault 1971, p. 347). However, as a discipline (defined as groups of objects, methods, propositions, techniques, and the like considered to be "true") matures, Foucault hypothesized

> that in every society the production of discourse is at once controlled, selected, organized, and redistributed according to a certain number of procedures, whose role is to avert its powers and its dangers, to cope with chance events, to evade its ponderous, awesome materiality. (p. 340)

Of the several sources of prohibition on "truth," Foucault (p. 343) viewed "the will to truth" as the greatest threat because establishment experts begin separating "true" discourse from "false."

Furthermore, laments Foucault, "The most superficial and obvious of the restrictive system [on discourse] is constituted by what we collectively refer to as ritual; ritual defines the qualifications of the speaker" (p. 350). Because a discipline seeks to control the discourse unduly by restricting who is qualified to speak, Foucault calls for a broadening of the legitimate range of "speakers" (p. 359). For example, in the case of economics, legitimate speakers include the "rich and poor, the wise and the ignorant, Protestants and Catholics, royal officials, merchants, and moralists." Freudian origins are apparent in his statement:

> We ask authors to answer for the unity of the works published in their names; we ask that they reveal, or at least display the hidden sense pervading their work; we ask them to reveal their personal lives, to account for their experiences, and the real story that gave birth to their writings. (p. 346)

Thus, we see in Foucault the outline of postmodern philosophy's emphasis on openness in discourse and radical democratization of authority. We also catch a glimpse of Nietzsche: the conspiracy by the elite to control intellectual disciplines–in Foucault's statement that:

> [Education] follows the well-trodden battle-lines of social conflict. Every educational system is a political means of maintaining or of modifying the appropriation of discourse, with the knowledge and the powers it carries with it. (p. 351)

*Jacques Derrida* (1930- ) Jacques Derrida is aligned with a dimension of postmodernism called *deconstructionism.* For Derrida, everything is text. He states (1982, p. 160) that the combining of language with the other threads of experience constitutes a "cloth." According to Derrida,

> the "strata" [of cloth or text] are "woven", their intercomplication is such that the warp cannot be distinguished from the woof....The discursive is related to the non-discursive, the linguistic "stratum" is intermixed with the prelinguistic "stratum" according to the regulated system of a kind of *text.* (p. 160)

Text must be *deconstructed* to reveal its meanings. Texts exist in a subjective context that favors one truth over another. There is no absolute truth; truth is relative to which person, discipline, or cultural perspective is writing or reviewing a text. Thus, texts must be interpreted in context and read in various dimensions to reveal their meaning. No one set of credentials or a perspective is accepted as authoritative. Although the subjectivity and relativism of deconstructionism give rise to many interpretations, its adherents contend that postmodern methodology is as "scientific" as analytical methodology–especially if science is defined as a framework of analysis rather than a set of propositions tested for their ability to predict.

*Postmodernism in Brief*

Postmodern philosophy by definition rejects grand narratives; hence, it has no creed or consensus articles. However, recurring themes are briefly summarized below.

1. Radical democracy and diversity in "truth" are prized in addressing socioeconomic and political issues. Expertise is accepted from a wide range of "authorities" having broad value perspectives. All is text or discourse and can be understood only in context. Science, art, and literature have numerous dimensions that can be understood only from their multifaceted perspectives. In science, for example, those perspectives come from humanities as well as sciences and from laypersons as well as professionals.
2. Power and authority control traditional knowledge. In western culture that authority flows from corporate business and profit that support the research, universities, faculty, communication systems, and elites who control disciplinary paradigms.
3. Subjectivity pervades all. Anglo-American analytical and postmodern philosophies alike recognize that neither science nor any other analytic human endeavor can be totally free of values, subjectivity, or bias. However, the analytical Anglo-American philosophy teaches that greater effort at objectivity is the proper response to excessive subjectivity in science. Postmodernism views objectivity as impossible and instead seeks accommodation with subjectivity in science. Because establishment science cannot be completely free from subjectivity, postmodernists argue that postmodern alternative science perspectives must counter conventional science. Examples where alternative perspectives are especially appropriate include income equity, gender and racial issues, environmental protection, and animal welfare. Compared to the analytical perspective, the postmodern perspective is more weighted to *normative* (value laden) than to *positive* (facts and logic) analysis.

As a result of these tendencies, the postmodern analyst tends to be more reactionary critic than creator. Perhaps because postmodernism holds no grand conceptual framework or utopian ideal, the philosophy tends to pessimism in its worldview and is positioned more for protest than progress.

## THE EMERGING INFLUENCE OF POSTMODERNISM

Postmodern philosophy was well developed by the 1960s, but had made little progress in displacing the dominant Anglo-American analytical philosophy. However, the intellectual community was increasingly disaffected by a system that in the twentieth century alone has allowed a great depression, two world wars, the Holocaust, colonialism, racial and gender discrimination, food insecurity, chronic poverty, and a Cold War.

If postmodernism needed validation, it seemed to receive it from the Vietnam War in the 1960s and 1970s. The West's involvement in the Vietnam War appeared to epitomize the failures of a modernist state devoted to the Enlightenment philosophy. Here, the "brightest and best" minds committed to rational thinking and with access to the best information and sophisticated analytical systems that science and technology could offer had blundered into a war it would not win. In contrast, a relatively disorganized band of less-informed young people (e.g., protesting college students) operating from nondoctrinaire "gut feelings" proved wiser, in their view, than the modernist forces of science, technology, and rationality. In the words of Bertens (1995, p. 5),

> postmodernism has been defined as the "attitude" of the 1960s subculture, or, somewhat more restrictively, as the "new sensibility" of the 1960s social and artistic avant-garde. This new sensibility is eclectic, it is radically democratic, and it rejects what it sees as the exclusivist and repressive character of liberal humanism and the institutions with which it identifies that humanism.

Armed with the surety of their philosophical system, postmodernists' disciples mounted their moral high horse, riding off with messianic zeal in all directions to do battle with the establishment. Many students who had protested the Vietnam War went on to earn academic credentials that led them to positions of intellectual influence. Even before the collapse of communism in the Soviet Union, postmodernism began to replace Marxism as the philosophy of choice especially among the disaffected who viewed problems of poverty, gender and race discrimination, and powerlessness as a product of modernism. Postmodernism also became the philosophy of choice among radical environmentalists who perceived that traditional knowledge is controlled by private businesses more interested in resource exploitation than environmental quality. On today's campuses postmodernist thinking is especially fashionable, though not necessarily dominant, in the liberal arts.

Postmodern philosophy, arising as it did out of the humanities and giving ascendancy to the complete metaphoric text, emphasizes that science must not be entrusted to physical scientists alone. It also follows, as noted earlier, that "agriculture is too important to be left to agriculturalists." The appropriate policy directions for agriculture are now debated by sociologists, anthropologists, philosophers, artists, journalists, environmentalists, and English teachers as well as by agriculturalists.

In contrast to the above ferment, economists have largely remained modernists wedded to neoclassical economics, positivism, and the Enlightenment. The neoclassical economics paradigm has remained resilient to every Kuhnian scientific revolution. It has preempted or incorporated appendages of Keynesianism and monetarism while largely deflecting inroads of socialism and Marxism. The seeds of conflict between moderns and postmoderns are indeed apparent, and are notable in the choice of ethical systems.

## ETHICS

Ethical issues can be divided broadly into (1) moral imperative and (2) utilitarian schools of thought. Postmodernists tend to subscribe to the moral imperative school and economists to the utilitarian school.

### Moral Imperative School

Moral imperatives (rights, obligations, rules) derive from two sources: (1) religious traditions such as the Bible or Koran and thus from divine authority, and (2) people. The Ten Commandments are the best known of an ethical system given by God.

Much more controversy surrounds ethical systems conceived by people. People may call their ethical systems human rights, animal rights, or natural law. Designating something a moral right or imperative implies that it cannot be denied to anyone under any circumstances or jurisdictions.

Acting alone, one individual could give little force or meaning to a right. Thus a moral imperative, for broad implementation, must have the support of many and be codified into public mores and laws. It follows that the moral imperative school may begin with personal conscience but ultimately is an exercise in persuasion and politics. Widespread political support and acceptance is most likely for a moral right that is consistent with existing divine and manmade laws, cheap to enforce, resonating with emotions, and responsive to the felt needs of the population. Moral imperatives as an ethical system are seriously flawed, however, in several ways.

One flaw is that moral rights have all the shortcomings of the political system. Pressure groups able to acquire emotional, money, or media support to further their rights agenda may not be representative of the interests of the public at large. Yet, their moral agenda of rights may be imposed on the public by the political system.

Rights can be very expensive to support. Traditional civil and political rights such as freedoms of speech, religion, and the press along with strictures against slavery, torture, and killing are not expensive to enforce because most call for government to do little. However, it has become fashionable in recent years to push for economic and social rights that can be financially costly if not prohibitive for a nation to supply. Examples include the right to food, health care, a living wage, a good education, and an end to poverty and economic inequality. These "rights" not only are expensive, they are difficult to define and enforce. Codification of these rights into law is a formula for endless litigation. Providing these public goods or safety nets may or may not have merit. But more sober judgments, thoughtful appraisal of issues, and sensible laws seem most likely to emanate from public decision makers unencumbered by the inflexible emotional baggage carried by the label "inalienable right".

The inherent inflexibility of moral rights hampers making of essential tradeoffs. No nation has sufficient resources to honor simultaneously the right to food, shelter, clothing, education, and health care for all. Moral imperative ethics provide no framework to make hard choices and complex compromises. One person's right can be another person's tyranny. Rights bestowed by the legislature on a farmer's animals may require expensive investments that bankrupt a farmer. "Thou shall not kill" makes sense except that innocent people must sometimes die to avoid death of an even larger number of innocent people. What if a "living" wage granted as a "right" to workers in a poor country is so high that it condemns them to poverty because firms are unable to hire workers or to invest profitably to create jobs?

Finally, many people believe that rights need to be attended by responsibilities. If able-bodied citizens do not have to work to obtain necessities, claims on the nation's resources to provide consumption goods and services may be so great that the nation cannot provide the savings and investment required for economic growth. The entire nation, including the poor, can suffer from such a policy. Thus, for able-bodied adults, the right to food may be attended by an obligation to work.

**Utilitarianism**

*Utilitarianism* is an ethical system that judges an action as desirable or undesirable based on whether it improves the well-being of people. Alternatively, utilitarianism has been defined as an ethical system that judges an action as right if it satisfies preferences (Randall 2002, p. 294). That definition invites criticism that utilitarian ethics sanctions the sadist who takes pleasure in seeing others suffer. The early utilitarians made clear that utility was of society and not just of individuals making choices in isolation.

As an ethical system, utilitarianism had origins in the work of Jeremy Bentham (1789) who defined *good* as pleasure and *evil* as pain. The moral person passed ethical judgment on actions according to whether they provided the greatest good for the greatest number of people.

That characterization has since been subject to intense criticism, mainly because it too seems to imply a self-centered hedonism of pleasure seeking if not orgy. John Stuart Mill (1806-1873) known for his seminal book *Utilitarianism* (1957), placed great emphasis on the multidimensionality of utilitarianism, noting how health, education, music, and the many simple pleasures of life can contribute to what might be variously referred to as utility, welfare, well-being, satisfaction, or contentment. Well-being does not necessarily imply aggrandizement of self or personal wealth; satisfaction for an individual may be enhanced by altruism toward other individuals or animals. Well-being is also influenced by trust in leaders and institutions, freedom, justice, opportunity for progress, and other intangibles, many of which are picked up by sociopsychological scales used to measure well-being.

**Synthesis**

Well-being is not just for the individual; it can apply to a collective of people and might include animals. That interpretation of utilitarianism raises troublesome questions of measuring satisfaction, of making interpersonal utility comparisons, and operationally implementing utility in policy on some objective basis.

For policy issues addressed in this book, utility of society is especially at issue so we can focus on group rather than on unreliable estimates of the well-being of individuals. Sociologists and psychologists have made great strides in quantifying the many dimensions of satisfaction. Using quantified sociopsychological scales, I have worked with colleagues to measure well-being of individuals and of factors contributing to it such as income, age, education, marital status, occupation, residence, and other variables (Blue and Tweeten 1997). Empirical results indicate that a person with very low income derives on average about 50 percent more satisfaction from another dollar of income than does someone with average income. And a person with average income typically derives about five times as much satisfaction from another dollar as someone with income five times the national average. Whereas such findings are difficult to apply in comparing the satisfactions of any two individuals, they suggest that a public safety net for the disadvantaged can add much to satisfactions in society but that a public policy transferring income from less to more wealthy groups is likely to reduce utility of society.

Neoclassical economics is built on the utilitarian principle that people act to increase their well-being. Economists predict behavior on the premise that individuals and firms will pursue actions providing them with private benefits in excess of private costs. And society is presumed to be better off by pursuing public and private ventures offering social benefits in excess of social costs. Thus, individuals and firms do not necessarily act in the public interest unless private incremental benefits (costs) are aligned with social benefits (costs) through taxes, subsidies, or other means for correcting such externalities. It follows that a role for public policy is to correct externalities so that incremental private costs (benefits) align with incremental social costs (benefits).

A criticism of the utilitarian ethical system used throughout this book is that it does not recognize that people do not always act in ways that are good for them or others (hence they need to be protected from themselves), that the majority sometimes wrongfully exploits the minority for the presumed "good" of society, and that utilitarianism is godless and ignores moral rights. These and other criticisms are addressed below.

So-called rules-utilitarians recognize that laws and rights can be utilitarian in promoting the general welfare. Utilitarian rules include protecting minorities against discrimination from majorities who out of meanness, sophistry, or ignorance judge that the public good is served by harming others. Rules-utilitarians recognize that laws and regulations on chemical use and highway safety are required to protect the welfare of people and animals from those who recklessly pursue their self-interests.

Utilitarianism need not be godless and ignore basic moral rights. I do not see a conflict between utilitarianism and religion–Christ made clear in the New Testament that the law was fulfilled in one new commandment–to love God and neighbor (Luke 10:27). The "neighbor" component of that commandment is very utilitarian. The Ten Commandments are not necessarily inconsistent with the command to love our neighbor. "Thou shalt nots" make clear to the dull or sophisticated that the welfare of society is reduced by bearing false witness or destroying the property and lives of others–however appealing the goals of racial or income equality being pursued. Such means are at once inconsistent with religious, humanitarian, and utilitarian ethical principles.

Neither is utilitarianism inconsistent with some of the better-known secular pronouncements on moral rights. Virginian George Mason, one of this nation's founders, is famous for authoring the 1776 Virginia Declaration of Rights, stating "that all men are by nature equally free and independent and have certain inherent rights...among which are the enjoyment of life and liberty, with the means of acquiring and possessing property, and pursuing and obtaining happiness and safety." Thomas Jefferson, in the famous Preamble to the Declaration of Independence, undoubtedly collaborated with Mason in writing that "all men are endowed by their Creator with certain inalienable rights, among which are life, liberty, and the pursuit of happiness." I interpret the above moral rights statements as ringing endorsement of rules-utilitarianism by persons committed to Enlightenment philosophy.

Implementing utilitarian ethics would appear to be impossible–who is to measure satisfaction resulting from each outcome in society to be certain that the well-being of society is enhanced? The answer is to measure the "rightness" of actions (means) rather than outcomes, and to look for systems known to foster utilitarian actions. Compelling evidence confirms that the market corrected for externalities and with a public or private safety net for those unable to provide for themselves is a proven instrument of utilitarianism (Tweeten 1999).

Economists like the utilitarian ethical system in part because it is consistent with a market economy. Adam Smith in 1776 wrote how individuals, though acting in their self-interests, were guided by the invisible hand of the market to achieve an end that was no part of their intention. That "end" is the well-being of society. In government, too, a check-and-balance system of separate branches of government, a free press, and the like are essential to turn private greed into public good. Such democratic-capitalism systems characterized by what I have called the *standard model* have far outperformed other systems in serving individuals and society (Tweeten 1999, pp. 473-78).

In contrast, moral rights ethics tend to lead to authoritarian socialist systems with a poor record of serving needs of people or the environment. While the standard model emphasizes private markets to determine what, when, where, and how to produce goods and services, it also emphasizes that a market will not function well without government providing public goods such as rules of the economic game along with a safety net for disadvantaged people. I will judge systems in subsequent chapters by the rules-utilitarian ethical yardstick.

## IDEOLOGY AND FINANCIAL SUPPORT FOR RADICALS

Postmodernism provides an ideology for radical agriculture; at issue here is the source of funding. Utilizing student and housewife volunteers minimizes funding needs for demonstrations.

Cash support is essential also. It is a perverse tribute to freedoms afforded by the nation's system of democratic capitalism that individuals and institutions made wealthy by that system are sometimes at the forefront of shameful efforts to peddle snake oil and violence. Ted Turner, a billionaire who made a fortune through free enterprise and globalization, allegedly is a source of funds for the Ruckus Society whose activists are trained to create mayhem at meetings of the World Trade Organization and other international groups. Willie Nelson, who also made (and lost) a fortune under free enterprise, gives Farm Aid concerts whose proceeds are used to undermine that enterprise system. (The reader is encouraged to examine funding sources on the website *consumerfreedom.com.)*

## CONCLUSIONS

Globally, postmodernism is replacing Marxism as the ideology of protest movements. Is a Kuhnian revolution under way, with the postmodernist philosophy of science about to replace analytical modernist philosophy in a new paradigm of science? Probably not, but it is helpful to remember that those schooled in the old tradition are the last to recognize the nascent revolution.

Distrust of authority and the establishment, visceral dislike for big corporate business, a general pessimism regarding science, and a call for radical democratization of "experts" by alienated and disenfranchised groups imply that postmodernism will continue to be prominent. However, it is unlikely that postmodernism per se can become the dominant philosophy of science, including economics, because at its core it is nihilistic. In rejecting grand narratives, it rejects itself as a coherent philosophy. Placed in a policy perspective, postmoderns cannot easily escape the profound contradiction of calling for more of an already big government they do not trust to right environmental, social, and economic wrongs documented by data they question.

In the logic of dialectical materialism, postmodernism is the *antithesis* of the Anglo-American analytical *thesis*. Out of the dialectical synthesis, an enriched new philosophy of science could emerge. Applied economists committed to the grand narrative of Adam Smith find themselves philosophically more distant from many of their colleagues in other social sciences and humanities. Without engagement through dialogue and multidisciplinary collaboration, many campuses will find themselves in the absurd position of having two entrenched philosophies of science–one for economics, business, physical sciences, and engineering, and another for postmodernists. Members of an animal species separated too long lose their ability to interbreed; truly interdisciplinary work on major socioeconomic and environmental problems will be impossible if dialogue among disciplines continues to be postponed.

Postmodern philosophy and moral imperative ethics help to explain how RAs think and act–their stress on gut feelings (emotions), moral relativism (rejection of absolute truth), their alienation, their use of propaganda to influence public opinion, and their appeal to government rather than to markets to right the wrongs of this world. Paradoxically, few RAs have heard the terms "postmodern philosophy" and "moral imperative ethics." However, many of the professors who teach postmodernist thinking to RAs as students are very aware of these ways of thinking. Thus, it is difficult to overestimate the importance of dialogue among intellectual leaders with various philosophies.

Applied economists who see a brighter future through science, technology, sound economics, and rational processes of democratic-capitalism can learn much from postmodernists. Modernist science that confines itself solely to positivism and statistical inference is incapable of addressing adequately many of the value-laden, seminal issues of the day. Postmodernism responds to problems that have not and cannot be addressed properly by double-blind experiments, quantification, computers, and hypothesis testing of traditional analytical sciences. Postmodernists have something to contribute by being sensitive to the potential bias in science arising from funding sources, from the personal values of the scientist, and from the values imbedded in the science; by being aware of the need to be inclusive with regard to minorities and the disadvantaged in general; and by being skeptical of controlling elites.

Scientists who solely adhere to the strict analytical rigor of modernism will not address many important problems of society, but throwing out science in favor of "gut feelings" and moral rights ethics is no solution. The science, technology, and market system that have been accused of getting us into some of today's problems also must be part of the solution.

Applied economists in their zeal to remain positivistic and objective have been preoccupied with economic efficiency. Meanwhile, many social scientists closer to postmodernism, radicalism, and populism have been preoccupied with equity (distribution of wealth and power) and have largely ignored economic efficiency. The gap between modernist economists and postmodern social scientists can be narrowed, however. For example, applied economists (Blue and Tweeten 1997) have shown that economists can use reasonably objective tools of statistics and other analysis to estimate a critical and inherently subjective parameter of equity, the marginal utility of income. That parameter enables economists to compare impacts on well-being among groups from alternative wealth distributions whether arising from changes in economic efficiency and/or income redistribution. While not satisfying the rigorous demands of purists, the analysis is probably considerably more reliable than a guess and can narrow the equity-efficiency debate. Another example is agricultural productivity gains from application of science favored by modernists but also consistent with postmodern goals such as lower food costs for poor consumers and saving wildlife and biodiversity by concentrating production on environmentally safe ground.

Finally, many ideologies in addition to postmodernism motivate radical and populist elements in agriculture. Economists like to assume that people act rationally and they like *Pareto optimal* policies that make people better off without making other people worse off. Such policies come along rarely if at all–almost any new policy makes someone worse off. So economists have reverted to favoring policies that raise efficiency (more output per unit of input) so that gainers potentially can compensate losers (in practice, compensation is rare) and leave all better off.

The world is not made up of rational people solely in search of monetary gain. Daniel Zizzo of Oxford University and Andrew Oswald of Warwick University in the UK found that the majority of people in their experiments were willing to lose money to put in place measures that reduced by even more the income of those they perceived to be financially better off ("The Economics of Envy," 2002, p. 69). Perhaps this explains some of the nihilistic behavior of radicals that economists consider to be irrational.

## NOTE

This chapter draws considerably on "The Challenge of Postmodernism to Applied Economics" (Tweeten and Zulauf 1999) with the gracious consent of the American Agricultural Economics Association and Blackwell Publishing, which holds the copyright. This permission does not necessarily imply any endorsement of the material in this chapter or book.

**REFERENCES**

Anderson, P. *The Origins of Postmodernity*. New York: Verso, 1998.

Bentham, J. *Introduction to the Principles of Morals and Legislation*. Amherst, NY: Prometheus Books, 1988. (Original publication 1789.)

Bertens, H. *The Idea of the Postmodern*. New York: Routledge, 1995.

Blue, N., and L. Tweeten. "The Estimation of Marginal Utility of Income for Application to Agricultural Policy Analysis." *Agricultural Economics* 16(1997): 155-69.

Deleuze, G., and F. Guattari. *Anti-Oedipus Capitalism and Schizophrenia*. Translated by R. Hurley, M. Seem, and H. Lane. Minneapolis: University of Minnesota Press, 1983 (original copyright 1972).

Derrida, J. *Margins of Philosophy*. Translated by A. Bass. Paris: Les Editions de Minuit, 1972; Chicago: University of Chicago Press, 1982.

"The Economics of Envy." *The Economist,* February 16, 2002, p. 69.

Foucault, M. *The Discourse on Language*. Paris: Gallimard, 1971. (English translation by R. Sawyer published in *Social Science Information.*)

Freud, S. *The Ego and the Id.* Translated by J. Riviere. Vienna: Internationaler Psychoanalytischer Verlag, 1923; New York: W.W. Norton & Company, Inc., 1960.

Ghebremedhin, T., and L. Tweeten. *Research Methods and Communication in the Social Sciences*. Westport, CN: Praeger, 1994.

Kierkegaard, S. *Philosophical Fragments*. Translation by David Swenson. 2d ed. Princeton, NJ: Princeton University Press, 1962.

Lyotard, J. *The Postmodern Condition*. Translated by G. Bennington and B. Massumi. Minneapolis: University of Minnesota Press, 1979.

Marx, K. *Das Kapital: A Critique of Political Economy.* Chicago: Henry Regnery Company, 1970.

Mill, J. S. *Utilitarianism.* New York: Macmillan, 1957.

Nietzsche, F. *Beyond Good and Evil.* Translated by R. J. Hollingsdale. New York: Penguin Books, 1973.

Randall, A. "Rational Policy Processes for a Pluralistic World." In *Agricultural Policy for the 21st Century,* edited by L. Tweeten and S. Thompson. Ames: Iowa State Press, 2002.

Tweeten, L. "The Economics of Global Food Security." *Review of Agricultural Economics* 21(1999): 473-88.

Tweeten, L., and C. Zulauf. "The Challenge of Postmodernism to Applied Economics." *American Journal of Agricultural Economics* 81(1999): 1166-72.

# 3

# Antiglobalists

## INTRODUCTION

Globalization refers to increasing worldwide cultural, economic, and political linkages (Diaz-Bonilla and Robinson 2001, p. 1). Globalization is the result of lower cost of communication and transportation, which in turn is the product of changing technology. The process of globalization is apparent in enlarged spatial flows of goods, services, technology, people, and ideas. The process brings the promise of a better life for many, but is unsettling to some people and institutions. This chapter examines autarky (no trade), opposition to multinational firms, and numerous other positions of antiglobalists. With few exceptions, those positions do not stand scrutiny. The final section of the chapter outlines elements of a policy for rich countries to better help poor countries, a principal stated objective of antiglobalists.

Globalization shapes the level and distribution of food security and wealth determining the well-being of people. A central question is whether the freer global flow of people and things makes the world more or less well off. Conventional agriculturalists (CAs) mostly answer yes. The antiglobalists among radical agriculturalists (RAs) mostly answer no to this question.

## GENESIS OF GLOBALIZATION

Globalization in human terms began thousands of years ago when early peoples left Africa for other lands. Eventually the migration reached the "ends" of the earth. Because of primitive transportation and communication technologies, isolation caused cultures, technologies, and living conditions to differ widely among regions and continents. Jerod Diamond (1997), in his magnificent panoramic tour of 12,000 years of human history over all continents, makes a powerful case that rising living standards coincided with the emergence of agriculture.

Development of agriculture began with the domestication of plants and animals. The consequent rise in food output permitted a major increase in population.

Increased agricultural productivity freed people from providing food to providing writing, numeracy, and literacy. Eventually, productivity gains allowed people to be educators, scientists, engineers, and the like–thereby fueling future growth in productivity and living standards.

Through trade and migration, globalization spread the fruits of advances in agriculture to the entire world. Cultures such as hunter-gatherer societies that resisted the advances unfortunately were overwhelmed by the better technology, disease control, administrative organization, and shear numbers of people attained by cultures that had the luck or pluck to change. The industrial revolution represented a milestone in globalization by accelerating improvements in transportation, communication, and other fruits of technology and science.

The "genie" of globalization is out of the "bottle," and it cannot be returned. Modern transportation and communication place virtually any material good or service in the hands of those who can pay. The fact that globalization moves us toward a "global village" causes much controversy. Communication satellites and portable radios and cassettes put American pop culture within reach of nearly every household. Even more controversial is that the $40 billion international arms market can place deadly force in the hands of almost anyone with a grievance to right.

Industrial nations of western Europe, North America, and Oceania spearhead globalization. Ironically, these industrial nations also provide most of the protestors against globalization. Tens of thousands of protestors, many of them RAs, met wherever leaders of the industrial nations (G-8), International Monetary Fund, World Bank, or World Trade Organization gathered in recent years. Many antiglobalist demonstrators were members of nongovernment organizations (NGOs) out of developed countries but working in developing countries, organized labor representatives, liberal-church clergy and lay-workers, environmentalists, and students and faculty from academic institutions (Kanbur 2001, p. 1).

Of the many thousands who protested at gatherings of world leaders, relatively few demonstrators were violent. Still, injuries occurred. For example, at the IMF-World Bank summit in Prague in September 2000, during ten hours of riots some 103 policemen were injured and 20 policemen had to be hospitalized (Ratnesar 2001, p. 34). In Genoa in July 2001 at a meeting of industrial (G-8) nation leaders, the antiglobalization agendas ranged from anticapitalism to "saving the earth to defending workers' rights and opposing free trade" (Ratnesar 2001, p. 33). One person was killed and many injured. The key issues of globalization were lost in the heat of confrontation and scattered violence.

The World Economic Forum, a discussion group of world "establishment" leaders from the public and private sectors, met in New York City in January of 2002 after meeting in previous years in Davos, Switzerland. Demonstrators were numerous, but violence was avoided because of tight security provided by New York City police. The World Economic Forum, like the Trilateral Commission, is viewed by some radical and populist organizations as the center of the international conspiracy to aggrandize banks and multinational corporations at the expense of people.

Meanwhile in January 2002, the World Social Forum, an antiestablishment counterpoint to the World Economic Forum, was meeting in Porto Alegre, Brazil. The World Social Forum recorded 51,000 delegates, three times the number meeting at the same forum at the same place in 2001 ("The World Social Forum" 2002, p. 32). Coincidental with that 2001 meeting, forum activists ransacked a McDonald's restaurant and genetically modified crops being grown on a Brazilian plantation owned by a multinational corporation. The star guest at the Social Forum in 2001 was Jose Bove, a French farmer who had served as a role model by earlier trashing a McDonalds restaurant and genetically modified crops in France. Delegates to the Social Forum mostly agreed that the IMF, World Bank, and World Trade Organization were villainous architects of world power, but despite their best efforts could not come up with alternatives to these institutions. Delegates were left supporting left-wing policies and institutions with a consistent record of failure.

## SEEING THE GOOD, THE BAD, AND THE UGLY IN GLOBALIZATION

Globalization homogenizes culture through modern communication, information, and transportation technology. Movies, television, books, music, and other pop-culture originating mostly from rich countries confront traditional religious beliefs and other dimensions of culture. I am not a fan of American pop culture, and sympathize with those who do not want to import it. Turning off the television set or radio is one solution. But neither antiglobalist demonstrators nor this chapter devote much attention to culture. Rather, the focus is on economic issues.

Proportions of people who live in poverty and food insecurity are lower than ever before, in large part due to globalization. Still, some 1.2 billion out of a global population of 6 billion people live in abject poverty defined as income of less than $1 per day (Hazel 2001, p. 1). Many of these people account for the approximately 800 million people who rarely get enough to eat and are classified as chronically food insecure (Tweeten 1999). Over half of the abject poor live in South Asia and sub-Saharan Africa.

Especially notable for this chapter is that three-fourths of the very poor live in rural areas where they depend on agriculture for their livelihoods (Hazel 2001, p. 1). Much of the globalization debate centers on how best to help these people. Of necessity, that debate must address agricultural issues, but problems of agriculture cannot be separated from issues of trade, multinational corporations, and economic development. This chapter makes the case that general poverty arises from lack of rather than because of globalization. This section is devoted to topics of appropriate economic policy, food self-sufficiency, and trickle-down growth.

### The Standard Model

Economics is central to issues of globalization. Redistribution of income will not alleviate global poverty. Income transfers from affluent nations are unreliable and will continue to meet only a small part of needs. Many poor counties have too small a "pie" of national income to redistribute. In other cases, redistribution either is politically impossible or economically undesirable because destruction of property rights would reduce national income. It follows that the pie of national income must be increased. With that pie to divide in now poor countries, a meaningful dialogue can take place on how to divide it.

General (as opposed to case) poverty is caused by failure to follow sound economic policy. The *standard model* (Tweeten et al. 1992; Tweeten and McClelland 1997, ch. 9; Tweeten 1999) has worked to raise national income and living standards wherever it has been applied. Any country can assure itself of the "pie" of income essential for food security and ending poverty (if not achieve affluence) by following the standard model. At the least, key parts of the model must be followed for success. Growth rates will depend not just on how well the model is followed, but also on the culture and natural resources of each country.[1]

The standard model emphasizes reliance on markets, privatization, and deregulation. The standard model is not an ideology, however, but focuses solely on what works for broad based development. Although the standard model calls for the market to make most decisions of what, when, where, and how to produce, the market only functions well if the government provides a supportive institutional environment. The few things that the government must do well for markets to function effectively include provision for the rule of law and for private property, sanctity of contracts, open trade, competition, balanced public budgets (except where borrowing has high payoff), sound monetary policy, key services such as primary schooling and agricultural research, and critical infrastructure (Tweeten and McClelland 1997).

A free press, independent judiciary, and democracy also can be very helpful. The free press and independent judiciary along with other checks-and-balances in government help to curb corruption. Democracy contributes to stability by resolving issues of succession in government.

A critical cornerstone of the standard model is *broad-based* development. That is, economic equity and efficiency are achieved simultaneously with proper public sector attention to investment in human capital and access to opportunity by all regions, races, genders, and creeds. Broad-based development provides a dividend to make transfers to those who cannot provide for themselves or find help from others. The height of the public safety net transfer system is a political issue to be decided by a (hopefully) democratic government–recognizing that a higher safety net may mean slower economic growth.

RAs take the opposite position from conventional agriculturists (CAs) on many if not most components of the standard model, but differences are most stark regarding the flow of trade, investment, and technology across boundaries. The CAs take the long view that the standard model has lifted hundreds of millions of people out of poverty and food insecurity. Recently, the World Bank studied 24 low-income "globalizing economies," including China and India, that are home to three billion people ("Is It at Risk?" 2002, p. 66). These countries substantially raised their trade-to-GDP ratios and more than coincidentally reduced their poverty rates while increasing their GDPs per capita on average by 5 percent annually in the 1990s. Meanwhile, GDP per capita increased "only" 2 percent per year in rich, industrialized countries.

Another two billion of the world's people live in low-income countries including Pakistan and much of Africa that became less globalized. In these countries, on average the trade-GDP ratio fell, GDP per capita declined 1 percent annually, and poverty rose in the 1990s (see "Is It at Risk 2002, p. 66). The World Bank study estimates that global free trade could lift 320 million persons out of poverty by 2015 to sharply lower food insecurity among the current 1.2 billion poor people in the world. Because an estimated 70 percent of exports of very poor counties is farm goods and textiles, it is clear that open trade has important implications for agriculture to be discussed later in this chapter.

Official development assistance (transfers) from rich to poor countries averaging about $50 billion per year is helpful, but is dwarfed by foreign direct-investment flows averaging about $240 billion annually to developing countries ("Is It at Risk?" 2002, pp. 66, 67). Most foreign investment moves among rich countries, but instructive exceptions exist. Ireland moved rapidly from one of Europe's poorest to one of its more wealthy countries by opening its economy–two-thirds of the country's output is now from foreign (multinational) firms. China lifted three hundred million people out of poverty by opening markets, thereby attracting foreign capital and multinational corporations. Thus, poor counties following sound economic policies attract private investment from abroad, causing their economies to bloom with attendant reductions in poverty and hunger.

Economic development has dwarfed the impact of transfers from government or from anyone else in lifting people out of poverty. Despite the central importance of economic development to better lives and environment in poor countries, numerous RAs and governments of poor countries steadfastly resist policy reform. Destructive economic policies are followed in part because persons in authority gain from those policies. Structural adjustment policies forcing countries to stop living beyond their means cause adjustment pains akin to a drug addict in withdrawal. When the very ill economic patient is brought into the doctor's office for treatment, the doctor is blamed for the inevitable recovery pain.

RAs, partly out of desperation from watching the casualties from structural adjustment "belt tightening" imposed on governments by the IMF and World Bank, pursue an agenda of the *right* to food aid, to local self-sufficiency in food, to low-input sustainable (if not organic) agriculture, to debt forgiveness, and to public as opposed to private enterprise. These deal with symptoms, not root causes. The challenge is to avoid the illness that brought the patient to the doctor. And the challenge is to cushion unfortunate side effects without giving up the cure.

Following the standard model avoids the *macroeconomic degradation process* (Tweeten 1999) attending the unwise policies of a nation attempting to live beyond its means. The degradation process is initiated when a nation attempts to consume more than it produces, borrow more than it can repay, and import more than it exports. For lack of other options after borrowing excessively at home and abroad, the next step is for the overextended country to create money to finance debt service and the government. Resulting inflation coupled with a fixed exchanged rate causes currency to be overvalued, thereby encouraging imports and discouraging exports. The ensuing shortage of foreign exchange, in the worst case, shuts down imports such as petrol and spare parts, bringing an economy to its knees. The poor suffer most from policies that depart from the standard model and energize the macroeconomic degradation process.

The important conclusion is that many if not most of the complaints of RAs and poor-country leaders ultimately trace to the hardships brought on the counties themselves because they reject sound economic policies. Numerous NGOs provide outstanding humanitarian services to poor counties at the local level but, tragically, follow the RA antiglobalist line on economic policy. NGOs have numerous contacts and hence influence in developing countries. These groups could become a constructive influence for sound economic policies in poor countries. Consequent economic development also can help the environment as explained in chapter 4.

**Regional Self-sufficiency and Export Cropping**

To promote economic efficiency and food security, many RAs call for regional self-sufficiency in food production and for an end to export cropping in poor countries. This proposal, like many other RA proposals, is counterproductive to RAs' stated goals. Subsistence agriculture, the ultimate in self-sufficiency, is characterized wherever it is practiced by abject poverty and frequent hunger.

Small, self-contained production regions have the greatest chance of production failure. The index of agricultural production failure (coefficient of variation, the standard deviation of annual production divided by the mean) averages only 1 percent for the world but for many local areas averages over 20 times this level. It follows that an area tied to world markets through trade need never experience hunger.

As a general rule, only poor people are food insecure. Except for armed conflict, people with income (resources) have been able to access food supplies that globally have been adequate to feed everyone every year since World War II. International markets constitute the most reliable food reserve. Nations such as North Korea that squander their national wealth and ability to earn foreign exchange in the quixotic pursuit of self-sufficiency and military might do not have sufficient buying power to enter international markets to cover local food production shortfalls and instead must rely on donations from other countries (see Sumner [2002] for South Korean analysis).

Cash cropping is a frequent target of antiglobalists' scorn. A study in Liberia found that a typical family could have three times as much rice, the local staple, and take better care of the environment by producing tree crops for export and purchasing rice in the market rather than producing rice for itself (Epplin and Musah 1987). Drawing on long-term research in Gambia, Guatemala, Kenya, the Philippines, and Rwanda, the International Food Policy Research Institute (IFPRI; von Braun 1989, p. 1) concluded that "agricultural commercialization raises the income of the rural poor, thus improving their food security."

Women in poor countries frequently produce food crops while men produce cash crops. Hence, the shift from food to cash crops could leave women worse off. However, the IFPRI study found, on average, net nutritional benefits to women and children as well as to men from cash cropping on all sizes of farms. The study also found that agricultural commercialization benefited hired workers engaged in food production, processing, and trading. This finding was especially gratifying because hired workers are some of the poorest in developing countries.[2]

Another RA charge is that regional self-sufficiency in food saves resources, particularly petroleum used in transporting food. Ohio, for example, could produce the coffee it imports from Brazil, the bananas it imports from Costa Rica, and the fruits and vegetables it imports from California. The cost in wasted resources and environmental degradation would be great as explained below.

The economists' version of St. Augustine's advice to "Love God and do as you please" is to "Price right and do as you please." Unfortunately, not all prices are "right" for people to promote the general welfare by responding to the market. Petroleum fuels are underpriced, not accounting for the full incremental social costs of pollution, global warming, highway congestion, resource depletion, and risky dependence on foreign oil. Ian Parry of Resources for the Future, a Washington, D.C., think tank, calculated that the average tax per gallon of gasoline, 40 cents per gallon in 2001, would need to double or triple to bring private cost of gasoline up to the social cost ("How Much Should Petrol be Taxed" 2001, p. 69). If prices were corrected, then firms maximizing profit and consumers maximizing satisfaction in response to market signals will properly conserve oil, soil, and other resources. Even if fuel prices were properly adjusted, Ohio consumers in all likelihood would continue to get their coffee, tea, and off-season fruits and vegetables from elsewhere.

**Does Economic Growth "Trickle Down"?**

International trade, like technology, allows more output from each nation's natural resources, thereby conserving resources and raising living standards. Technology is a greater force than trade to create more with less, but receives less criticism from RAs.

William Cline of the Institute for International Economics estimated that trade accounted for only 7 percent of the increasing ratio of skilled to unskilled workers' wages from 1973 to 1993, or one-fifth as much as technology (Crook 2001, p. 9). Should we therefore halt trade and technological change to reduce income inequality? The issue cannot be separated from the larger issue of "trickle down," which has troubled RAs for decades.

In industrialized countries, economic growth fostered by foreign fixed investment, technology transfer, trade, and other dimensions of globalization favors capital-intensive, high-wage industries such as information technology and pop culture. This pattern would appear to raise inequality, with growth bypassing unskilled workers. However, wages in less-favored occupations are bid up by competition for workers in occupations favored by economic growth. Economic growth provides resources to invest in schooling and skill training for low-income and other workers, raising their earning capabilities. Productivity growth from technology or trade raises the value of a nation's currency, thereby raising buying power of consumers across the board. For the above reasons, less-skilled as well as more-skilled workers in industrial countries earn more than workers in developing countries.

In poor countries, globalization raises international demand for labor-intensive goods and services produced mostly by low-wage, low-skilled workers. Their earnings increase as they shift employment from traditional occupations to export industries. As domestic and foreign-owned firms bid for more workers and more skills, wages rise. Some of the dividend from development is invested in human capital, further raising personal earnings. A recent World Bank study of 80 countries found that income of poor people increased as fast as income of others with economic growth (Crook 2001, p. 10). Economic development does indeed trickle down to benefit the poor.

Hundreds of millions of people have been lifted out of poverty and food insecurity in recent decades by economic growth. Most notable examples are the Tigers of Asia: Taiwan, South Korea, Hong Kong, and Singapore. Other countries such as China also have made great progress as noted earlier. That success of countries in reducing poverty and food insecurity is closely related to how fully they have accepted globalization and other parts of the standard model.

Global numbers of chronically food insecure people fell from 917 million in 1969-71 to 839 million in 1990-92 and are projected by FAO to fall to 680 million by year 2010 (FAO 1996). Only one region of the world, sub-Saharan Africa, seems to be losing the capacity to feed itself. The proportion of that region's people projected to be food insecure in 2010, 30 percent, is more than double the proportion in any other region (see Box 3.1). Sub-Saharan Africa of all regions has been least globalized and least multinationalized. Some countries such as North Korea–whose residents would be starving in the absence of food aid–are even more isolated from international commerce and investment. But that example reaffirms the conclusion that triumph of antiglobalization forces would relegate additional millions of poor people to chronic food insecurity.

## INSTITUTIONS IN GLOBALIZATION

Institutions play a critical role in economic development and have received their share of criticism by RAs. Antiglobalists charge that a few powerful governments and multinational corporations control the international economy. These powerful actors headquartered in rich nations allegedly aggrandize themselves on the backs of poor people and poor nations. Economic growth is said to be a zero-sum game, with progress among the rich nations coming from immiserization of poor nations.

RAs allege with some justification that rich countries block the path to economic betterment for poor countries. That path normally progresses successively from reliance on agriculture, to food processing, to light manufacturing, and finally, in later stages of development, to high-tech industry and services. Rich countries, their farmers subsidized by government, dump agricultural products on markets of poor countries to the detriment of local producers. Developing countries have difficulty progressing beyond primary agriculture because rich countries impose tariffs and quotas on imports of processed foods, textiles, footwear, and other high-order but labor-intensive products and services from developing countries.

**Box 3.1**
In 1988, I served on a team of three American and three Tanzanian social scientists to determine "the root cause of third-world socioeconomic injustice" in Tanzania. The team, sponsored by a major liberal Christian organization, was given that assignment and asked to test three hypotheses: that the root cause was (1) multinational corporations, (2) neocolonialism, and/or (3) declining terms of trade apparent in falling prices for exports relative to imports of Tanzania.

Each of the team members had considerable experience in Tanzania, and additional analyses were done as necessary for a competent overall study. The American team reported back that none of the hypotheses explained the general poverty found in Tanzania. Outside forces (as implied by the three hypotheses) did not bring Tanzania from one of the more prosperous sub-Saharan countries at independence to economic ruin by 1988. Rather, we reported that the economic ruin of Tanzania was caused by the country's own policies (or more accurately, Julius Nyerere and his African socialism policy called *Ujama*). Uncaring outsiders also could not be blamed–Tanzania enjoyed record economic and humanitarian assistance from abroad during its descent into the economic abyss. Our finding of a self-inflicted policy wound was not well received by church officials–we were fired immediately after giving our report.

Tanzanian terms of trade and exports had indeed fallen. The drop could be explained by the high export taxes imposed by the Tanzanian government on coffee and the failure of government to invest in research to improve agricultural productivity. The lesson is that an economy cannot ignore sound economic policy without being bypassed by the rest of the world.

Antiglobalists claim that public and private financial institutions in rich countries, the IMF, and World Bank make credit too readily available to poor countries. Such countries have a high preference for present consumption. It is said that poor countries accumulate debt because they lack capacity to screen and reject credit for projects offering low economic payoffs. A large portion of exports must be devoted to debt service, compromising the ability of poor countries to pay for food, petrol, spare parts, and other inputs critical to their economy and the well-being of poor people.

This section is directed especially at controversial institutions of trade and development, including multinational corporations and the World Trade Organization. I make the case that the institutions in some instances need reform but on the whole have performed well. Emasculating multinational corporations and the World Trade Organization, as proposed by antiglobalists, would hurt the interests of poor countries and rich countries alike.

**Multinational Corporations**

RAs view multinational corporations as the panzer divisions of international capitalism. They presume that corporations wield power to control culture and knowledge at the expense of local culture and individual thought. RAs have noted that a private firm lacking competition and facing externalities such as open access property can pollute the environment, turning private greed into public harm. Multinational corporations have been faulted for exploiting workers, moving jobs from high- to low-wage countries, and for despoiling the environment.

Many such charges lack merit. Multinational firms bring poor countries much needed capital, technology, management, and access to international markets. Many poor countries do not follow essential elements of the standard model, hence do not attractive foreign investment. Most foreign direct investment (FDI) goes from wealthy countries to wealthy countries. While RAs lament the heavy hand of multinationals in poor countries, poor countries have too little rather than too much investment by multinationals and desperately want and need more.

Multinational firms engaged in manufacturing on average paid 50 percent higher wages than domestic manufacturing firms in 1994. (Data are from Edward Graham of the Institute for International Economics as reported by Crook [2001, p. 13].) In low-income countries multinational firms paid double the wage of local firms. Separate studies for numerous individual countries confirm these results, including those for Mexico showing wages are highest near the border with the United States where multinational firms are ubiquitous. The complaint I have most frequently received from farmers in poor countries is that multinationals pay such high wages that no hired workers are available for farm work at an affordable wage.

Another charge is that the intense international competition among countries for multinational firm investments and job creation drives down the taxes that governments can extract from multinationals. Consequently, taxes are not available to finance essential safety-net programs. This charge is easily refuted. The Scandinavian economies are at once some of the most open and the most taxed in the world, with government receipts accounting for over half their Gross Domestic Product. Funds for human resource development, agricultural research, infrastructure, and the welfare state seem to be plentiful in these open economies.

Antiglobalists also claim that many multinational companies are larger and more powerful than governments, hence can control the fate of nations as well as firms. Part of this misconception stems from inappropriately comparing *gross* revenue of corporations with *net* revenue or product of governments. Multinationals, unlike governments, do not have power to tax, raise armies, pass laws, or send people to prison. Even large multinational companies are occasionally thrown out of countries–seemingly more often out of small, poor countries than out of rich, large countries.

The option of firms and people to leave countries can constrain bad behavior of governments. Globalization enhances alternatives for firms, placing pressure on governments to respect multinationals *and* their own citizens. The standard model ("golden straight jacket" in the terminology of Thomas Friedman [2000]) constrains policies of governments if they wish to see their citizens prosper. Given the wide range of options for greater well-being of people that can be bought with the fruits of following the standard model, an electorate that binds its elected officials to that model may be quite rational. Consumers will enjoy greater variety *and* quality of consumer goods along with more buying power in an open economy.

Antiglobalists blame globalization for corporate monopoly exploitation, but improved transportation, communication, and open markets create more, not less, competition. Sometimes a corrupt local government will allow only one multinational firm to locate. The exploitation of consumers that follows is due at least as much to government corruption as to the greed of the multinational. An alternative is autarky, but exploitation by a local, domestic monopoly can be especially pernicious. Not only is there lack of price competition, there is likely to be no international press or other body calling for reform in poor countries especially prone to cronyism, a controlled press, and weak public antitrust institutions.

**Free Trade versus Fair Trade**

"Free trade" may be defined as trade among nations unencumbered by barriers except as allowed by the World Trade Organization. "Fair trade," promoted by many RAs, is defined as trade consistent with protection of workers, women, minorities, children, sustainable agriculture, and the environment. Free trade leaves most social and environmental issues to be decided by each individual country. Fair trade would impose social and environmental rules or "chapters" on all countries, rich and poor, before markets would be opened.

Multinational firms sometimes exploit the environment in developing countries; so do local firms. Economic development through exploiting comparative advantage can help poor countries to afford public goods such as protection for forests, land, air, and water. The comparative advantage of poor countries lies in labor-intensive industries made competitive by abundant, low-wage labor. What international agency bureaucrat is competent to judge what is a "fair" or "just" wage for poor countries? A wage set too low would not raise wages. More likely, the wage would be set so high that investors would not create jobs and workers would continue to have few job opportunities and receive low earnings.

Laborers work at multinational firms because alternative employment is inferior. Policies requiring multinationals to pay high wages cut employment opportunities by removing the one advantage poor counties can offer multinationals–low-cost labor. If workers are being exploited–paid less than their contribution to the value of output–count on "greedy" competitors to attempt to raise profit by bidding away such workers with higher wages.

International trade rules that set wage standards enforced by terminating imports from noncomplying poor countries "stiff" consumers in developed countries by raising the cost of imports, and "stiff" workers in poor countries by cutting employment and wage opportunities. It is not surprising that support for imposing labor and environmental standards on poor countries comes from self-serving labor unions and environmental groups in rich countries.

It is unclear, however, why liberal church groups and NGOs that purport to care about poor people favor the labor and environmental chapters in international trade agreements that poor countries oppose. Workers in developing countries rightly question why well-intentioned people in rich countries condemn them to poverty and starvation.

While many objectives of "fair trade" are commendable, the pursuit of unobtainable "fair trade" by RAs drives out attainable "free trade." It is notable that efforts to begin a new round of multilateral trade negotiations in Seattle in late 1999 broke up when none of the some 100 poor countries among the over 140 countries in the World Trade Organization would support labor and environmental chapters in a trade negotiating agenda. Thus, "fair trade" ultimately is autarky.

### WTO a Supranational Tyrant?

The World Trade Organization has been faulted by antiglobalists on the right and left for being an unelected world government dictating policies to sovereign states–even to powerful entities such as the United States and European Union. In fact, the WTO and its set of rules originated and exist with unanimous approval by the more than 140 member countries. The purpose is to avoid the so-called prisoners' dilemma –countries pursuing policies that initially make each individually better off but, when widely followed, make them collectively worse off. Although the world and each country are worse off when each country protects its market, no one country thinks that it can drop that globally disastrous policy without losing.

While the premise that unilateral liberalization hurts the liberalizing country is questionable, nonetheless, countries are more willing to give up welfare-reducing trade distortions if other countries do the same. Countries are free to leave WTO, but more are choosing to join than to leave. In short, WTO can be criticized with greater justification for being a paper tiger than for being a dictatorial lion.

## BENEFITS OF MORE OPEN TRADE

Intense opposition by antiglobalists to free trade prompts a closer look at potential benefits from more open markets. Many antiglobalists argue that they are for free trade and would support it when they see it. Of course, the world will never have completely free trade or even what antiglobalists call "fair" trade so the above position is tantamount to isolationism. This section makes the case for unilateral free trade. It pays a country to reduce its trade barriers if other countries do not reduce theirs. Many-country (multilateral) free trade agreements are even better.

The Canada-U.S. Trade Agreement (CUSTA) of 1989 paved the way to include Mexico with the United States and Canada in the North American Free Trade Agreement (NAFTA) signed in 1993 and implemented in 1994. Mexico's trade liberalization had begun in earnest with its accession to the General Agreement on Tariffs and Trade in the mid-1980s but trade picked up even more with joining of NAFTA. Tweeten, Sharples, and Evers-Smith (1997) estimated that CUSTA/NAFTA added $1.4 billion to U.S. agricultural exports to Canada and $1.9 billion to Canadian agricultural exports to the United States by year 1995 over 1989 exports.

Burfisher, Robinson, and Thierfelder (1998, p. 66) found lower benefits from NAFTA, but all estimates indicate that NAFTA created considerable trade.

Canada and Mexico each accounted for only 5.3 percent of U.S. farm exports in 1989 (table 3.1). That share nearly tripled by 2001, despite the growth-retarding Mexican peso crisis in 1995. Under NAFTA, Canada and Mexico have displaced the European Union and are rapidly overtaking Japan as the largest U.S. farm export markets.

**Table 3.1. U.S. Agricultural Export Shares by Selected Country (Region) for 1989, 1994, and 2001**

| Region or country | 1989 | 1994 | 2001 |
|---|---|---|---|
| | Percentage of United States | | |
| Canada | 5.3 | 12.5 | 15.0 |
| Mexico | 5.3 | 9.2 | 14.0 |
| European Union (EC-12) | 19.2 | 15.9 | 11.6 |
| Japan | 46.2 | 21.6 | 16.8 |
| United States | 100.0 | 100.0 | 100.0 |
| $billion | 38.0 | 42.5 | 53.5 |

*Source:* U.S. Department of Agriculture (October 2001, p. 47, and earlier issues).

Gehlhar (1998, p. 40) estimated that NAFTA added $2.3 billion annually to U.S. national income compared to $4.6 billion added by multilateral trade liberalization of the Uruguay Round completed in 1994. Long-term economic benefits of trade are much greater, as we shall observe later.

Trade has little effect on the overall employment rate. If it did, countries such as Singapore and Luxembourg that trade heavily would have massive unemployment. Rather, trade creates *better* jobs and adds to real national income. One "cost" of more open trade is job shifts, which for disadvantaged workers can be traumatic although overall job quality and remuneration rise on average. Thus programs are useful to aid adjustment to other employment of workers displaced by trade.

A Free Trade Area of the American (FTAA) by year 2005 was endorsed in the first hemispheric summit in 1994 in Miami and reaffirmed in Santiago, Chile, in 1998. According to Gehlhar (1998), the western hemisphere FTAA if implemented could raise income of the United States by $3.3 billion per year. Poor counties in the hemisphere have relatively the most to gain.

### Gains from Ending Global Agricultural Trade Distortions

The United States especially can benefit from multilateral freer trade in agriculture in part because our agricultural tariffs average 12 percent (Burfisher 2001, p. 10) compared to a world average of 62 percent (Gibson et al. 2001). American agricultural product tariffs average less than those of our major farm markets: Canada, 24 percent; the EU, 21 percent; and Japan, 33 percent (Wanio, Gibson, and Whitley 2001).

As shown in table 3.2, agricultural market distortions such as tariffs and subsidies reduce global annual real income by $31 billion in the short run to $56 billion in the long run–the latter accounting for cumulative benefits from greater savings, investment, and productivity (Diao , Somwaru, and Roe 2001). Estimated benefits to the U.S. from agricultural free trade totaling $6.6 billion in the short-run (static) scenario constitute 21 percent of global benefits. Benefits to the United States in the long run–recognizing direct and indirect impacts on saving, investment, and productivity–total $13.3 billion in the fifteenth year after liberalization, or 24 percent of world benefits. If annual benefits of that level are maintained in perpetuity and discounted at 5 percent, the present value of all future benefits of free agricultural trade total $266 billion to the United States and $1.13 trillion to the world.

Although a few developing countries are worse off with free agricultural trade because they are food importers who will pay higher prices for imports with liberalization, gains far exceed losses for most countries (table 3.2). Thus, the foundation for an agreement providing freer trade appears to exist.

A general rule is that the principal costs of trade distortions are borne by the countries that practice them (Tweeten 1992, p. 279). This principle is supported by data in table 3.3 indicating the impact on world agriculture commodity prices from eliminating policy distortions. Distortions are divided into three types: (1) tariffs, (2) domestic supports such as commodity price supports and supply management, and (3) export subsidies. Together, the European Union and United States accounted for 51 percent of the static and 42 percent of the long-term dynamic potential gains from trade. The two entities together accounted for 52 percent of the world price distortions reported in table 3.3.

Multilateral markets liberalization, according to Diao et al. (2001), potentially could raise world agricultural trade prices by 11.6 percent (table 3.3). Of this increase, half comes from eliminating tariffs (especially used by importers, notably in developing countries), nearly one-third comes from eliminating domestic supports especially prominent (80 percent) in the United States and European Union, and nearly one-sixth comes from ending export subsidies especially prominent in the EU (table 3.3).

With few exceptions, countries, regions, and the world gain under either unilateral or multilateral trade (and commodity program) liberalization (Makki , Tweeten, and Gleckler 1994).[3] Producers gain more (or lose less) for every country or region from multilateral than from unilateral liberalization (Rodney Tyers and Kym Anderson in World Bank 1986, pp. 128-31; Makki et al. 1994). Thus, producers have a major stake in success of multilateral liberalization.

In the long run, full liberalization could increase the real value of U.S. agricultural exports by 19 percent and agricultural imports by 9 percent according to Burfisher (2001, p. 7), hence would markedly improve the nation's balance of payments. At issue is how such gains would be divided.

**Table 3.2. Annual Welfare Impacts from Elimination of Global Agricultural Tariffs and Subsidies**

| | Static resource allocation gains[a] | Static plus dynamic benefits from investment growth plus productivity gains[b] |
|---|---|---|
| | Billion 1997 U.S. Dollars | |
| **World** | **31.1** | **56.4** |
| **Developed country group** | **28.5** | **35.2** |
| Australia and New Zealand | 1.6 | 3.5 |
| Canada | 0.8 | 1.4 |
| EFTA | 1.7 | 0.2 |
| European Union | 9.3 | 10.6 |
| Japan and Korea | 8.6 | 6.2 |
| United States | 6.6 | 13.3 |
| **Emerging and developing country group** | **2.6** | **21.3** |
| China | 0.4 | 2.2 |
| Latin America | 3.7 | 6.1 |
| Mexico | -0.2 | 1.6 |
| Other Asian countries | 1.5 | 5.1 |
| South African countries | 0.3 | 0.8 |
| Rest of world | -3.1 | 5.4 |

*Source:* Diao et al. (2001)

[a]Static gains refer to the annual gains due to removing distortions to production and consumption decisions.

[b]Dynamic gains include effects related to cumulative increases in savings, investment, and productivity over a 15-year postreform period. Dynamic welfare impacts are the annual level about 15 years after reform.

**Table 3.3. Effects on World Agricultural Prices of Eliminating Agricultural Policy Distortions, by Country and Policy**

| Elimination of: | World[a] | U.S. | EU | Japan/Korea | LDCs |
|---|---|---|---|---|---|
| | (Percent change from base price) | | | | |
| All policies | 11.6 | 1.8 | 4.4 | 1.5 | 2.3 |
| Tariffs | 6.0 | 0.7 | 1.5 | 1.4 | 2.3 |
| Domestic support | 3.6 | 0.9 | 2.0 | 0.2 | b |
| Export support | 1.5 | 0.1 | 0.9 | b | 0.0 |

*Source*: Diao et al. (2001)

[a]Numbers do not sum to row and column totals because only selected countries are included and there are interactions among policies.

[b]Not applicable, no policy in use.

Diao et al. (2001) estimated gains in world prices by commodity resulting from elimination of all policy distortions. Gains for wheat (18.1 percent), other grains (except rice, 15.2 percent), sugar (16.4 percent), and livestock products (22.3 percent) exceed the world price average gain of 11.6 percent. Gains are especially large for sugar and livestock prices with global tariff removal, and in wheat and other grains with developed country (OECD) subsidy removal. Gains are less for individual commodities from global export subsidy removal. U.S. major crops (except fruits and vegetables) and livestock could benefit from higher world prices associated with multilateral trade and commodity program liberalization.

Other things equal, an 11.6 percent increase in the price of agricultural commodities would raise farming receipts averaging $193.2 billion annually for 1998-2000 by $22.4 billion. In the same three-year period, that potential gain exceeds the annual value of all direct payments (including loan deficiency payments) averaging $16.0 billion or Commodity Credit Corporation outlays averaging $20.5 billion (U.S. Department of Agriculture, June/July 2001, p. 58). Thus, the economic gains to U.S. farmers from freer trade potentially could compensate producers for government payment losses from commodity program (and trade) liberalization.

## SUMMARY, CONCLUSIONS, AND POLICY RECOMMENDATIONS

Globalization offers proven, massive social and economic advantages to rich and poor countries alike. Long-term global benefits from free trade in agricultural products alone total over $1 trillion. Freer trade raising income also reduces poverty, hunger, and environmental degradation. Women, children, and minorities tend to be treated better in countries with higher incomes per capita, especially in countries that pursue the broad-based growth strategy of the standard model. Antiglobalists' prescriptions, if implemented, would deny benefits of freer trade. Many of the antiglobalists' grievances are over the relationship between rich and poor countries. As such, many of the following policy suggestions give advice on how best to respond to those grievances, and to secure more of the potential benefits of globalization.

*Rely on markets to guide globalization.* Given that globalization is the result of modern transportation and communication technology that cannot be turned back, the issue is how to guide globalization. Even antiglobalists mostly support globalism in the form of information technology, a universal declaration of human rights, the right to food, and labor and environmental chapters in world trade agreements. For many activists, globalization is a code word for the worldwide hegemony of American and Anglo-Saxon culture and institutions. In keeping with the nihilist bent of postmodernism, however, antiglobalists offer no constructive alternatives to replace democratic capitalism.There are two broad means to direct the tide of institutions and culture. One is by government decree, with an elite few dictating what is good for the public. McDonald's and Monsanto, for example, surely would by kept out of many countries if antiglobalists were in charge. A second means of control is to rely on markets. Those who dislike a product, firm, or technology need not use it. If enough people fail to buy, the product and firm would withdraw for lack of profit. On the other hand, favored products and firms would prosper whether domestic or foreign. The market approach to globalization requires the usual public measures to protect food safety and the environment, but offers the obvious advantage of freedom of choice prized by consumers everywhere.

*Provide more development assistance to poor countries, but provide no more than humanitarian assistance to any poor country that will not reform unsound economic policies.* Every country has sufficient natural or human resources to end domestic poverty and hunger. Every country has available to it a proven economic framework, the standard model, to end poverty and hunger. Failure to follow the model traces to politics, which in turn traces to culture (including tribalism) and economic illiteracy. Having proper policies in place is essential before economic assistance can work. The World Bank and other multilateral institutions have not had much success in using aid to leverage reforms. Thus, the Meltzer Commission proposal, that aid (except humanitarian assistance) be given to countries only *after* they reform their policies, has merit.

Whereas developed nations can do much to open markets and develop technology for poor countries, in the final analysis poor countries themselves will have to make the policy decisions required for their people to escape economic and social deprivation. Following antiglobalists' nihilistic nostrums would needlessly condemn millions of people to hunger and poverty.

*Industrialized countries can do most to help poor countries by opening markets.* Barriers of rich counties are especially large to imports of processed foods and labor-intensive goods from poor countries. Of course, poor countries will benefit themselves and other countries by reciprocating–opening their own markets.

Benefits tend to be large from unilateral or multilateral liberalization. However, agricultural producers in most countries receive more income with no liberalization than with unilateral or multilateral liberalization. Consumers far outnumber producers in every country, hence the losses per producer tend to be larger than gains per consumer with liberalization. Each producer as a big loser is motivated to organize with others for political action to stop reform. Those producers tend to be more than a match for large numbers of complacent consumers in the political arena *even though overall gains to consumers far exceed losses to producers from liberalization.* Job losses from trade are easily visible while gains from trade, though larger than losses, are not so visible because they are highly disbursed. Thus, market distortions remain because they benefit some producers and workers, but consumers, taxpayers, and the public at large are better off with liberalization.

Compensation by taxpayers and consumers for losses to producers can help to resolve the welfare-reducing strategic behavior of producers and workers described above. Given the net economic benefits (deadweight gains) available from liberalization, compensation can be provided to farmers and laborers to make them as well or better off with than without liberalization while leaving a net benefit for the rest of society.

Such compensation to farmers may be essential to liberalize trade and commodity policy in the short and intermediate run. The Federal Agricultural Improvement and Reform (FAIR) Act of 1996 and related policy changes in the 1990s brought fundamental reforms in the United States compatible with freer domestic and foreign markets. Chief among these were a shift from coupled deficiency payments to decoupled payments, an end to set aside (supply management), and less engagement of government in commodity stock accumulation and export subsidies.

The 2002 U.S. farm bill returns the United States to failed policies of the 1950s that featured generous economic support from taxpayers in the absence of production controls. Excess production generated thereby will be dumped in international markets. It is hypocritical of the country to dump agricultural products on foreign markets to the detriment especially of poor-country producers while we apply strong antidumping penalties on imports of steel and lumber allegedly dumped on our market.

Unilateral or multilateral liberalization of trade could be attended by reliance on direct, decoupled payments for all farm commodities. These payments do not raise income of farm operators except in the short run, hence could be phased out over (say) five years. Unilateral termination of commodity programs including direct payments totaling 42 percent of net cash farm income in year 2000 would appear to be traumatic to producers. However, reduction of transition payments could be offset (for farm income) by rising farm commodity prices and receipts resulting from (1) less farm output attending lower loan rates and crop insurance subsidies, and (2) world farm commodity price-enhancement from freer global trade.

*Industrialized countries can promote international development and equity by providing high payoff public goods that poor countries will not provide for themselves.* Contrary to RA thinking, trade between affluent and poor countries is a positive-sum game benefiting both sets of countries. Low-income countries make poor trading partners because they lack buying power. An attractive alternative to promote development is for wealthy countries to invest in basic research–poor counties cannot afford to do so. Public research in developed countries can develop technologies made available globally and at low cost to poor countries. Basic research on crop and animal genetics to cope with stress from disease, insects, drought, cold, and heat offers high economic payoffs to rich and poor countries.

The 16 institutions comprising the Consultative Group for International Agricultural Research are located at key locations worldwide and provide critical technology, especially for poor countries. Funding has been flat for many years. Rich countries can do more to fund such agricultural research. It is a critical bridge between basic and applied temperate zone research in industrialized countries on the one hand, and, on the other hand, applied adaptive research essential to tailor emerging technologies to use in poor tropical and subtropical countries.

*Investment in human capital, like basic research, offers high economic payoff to poor countries and can be performed in developed countries.* Financing from developed countries can help persons from poor countries study technology, management, and policy in economically successful countries. Incentives can encourage students to return home to apply their knowledge to raise living standards. In other cases, through distance learning technology or other means, developed countries can assist learning *within* poor countries.

*Industrialized countries maintaining appropriate macroeconomic and commodity program policies create a more stable world economy.* As indicated earlier, poor countries tend to prosper when other countries prosper and vice versa. Sound macroeconomic and farm commodity program policies in the United States and elsewhere that follow the standard model, let markets work, and minimize business and commodity cycles can help developing countries directly and also can set a good example.

*Poor countries need to be allowed to file for bankruptcy.* Individuals file for bankruptcy when debts overwhelm, and they need a new start to get on with their lives. Should poor countries be denied the same recourse? Much poor-country debt is uncollectable and might as well be written off. In many cases wealthy countries pushed it on poor countries that made bad investments because they were not good at judging how to wisely spend or to repay loans.

No institutional framework exists to administer country bankruptcy comparable to a court overseeing individual bankruptcy. Anne Krueger of the International Monetary Fund has proposed a framework in which the fund would organize key creditors to work with debtors in writing off or restructuring debt. Some of the following proposals would reduce the frequency of countries experiencing financial difficulty so that bankruptcy need be employed sparingly.

*Provide development assistance to poor countries as grants, not loans, and phase out the World Bank.* The World Bank has little to show for the some $500 billion of loans made in the past half-century (Bishop 2002, p. 28). Barely one-third of its projects in poor countries have achieved acceptable results by its own review. It is not surprising that many countries cannot repay the loans foisted on them. The World Bank has outlived its usefulness. Private lenders will be more prudent in making loans and have plenty of funds available for countries following sound economic policies.

Official development assistance given as grants will not become uncollectable debt. Private lenders and investors will be less inclined than the World Bank to finance schemes for political reasons. Private lenders have expertise to scrutinize potential clients for their ability to service debt. Massive private capital seeking profitable places to invest constitutes a major incentive for poor countries to curb violence and corruption and to embrace sound economic policies.

*Private investment can be compatible with preserving the environment.* For example, the most economically attractive place to expand timber production is in transition zones between rain forests and more arid lands. The fast timber growth at low cost and attractive profit of plantation forests in transition areas will drive down timber prices to make logging of ecologically fragile but biodiverse rain forests economically unattractive. Similarly, private and public investment in bio- and other technology to raise yields has freed millions of hectares for grass, trees, and biodiversity that otherwise would be in soil-depleting crops. Investments to foster continued technological progress need to be encouraged.

*International rules and institutions of globalization could use some reform.* Highly mobile short-term financial capital moves so quickly in and out of countries that serious financial crises sometimes result. Countries are experimenting with means to slow the flow and stabilize economies over the business cycle. Chile provides an example for a policy encouraging longer-term capital inflows and discouraging short-term financial portfolio flows.

In conclusion, the importance of globalization cannot be overemphasized. It constitutes the best and probably the only hope for ending the poverty and food insecurity caused by misguided policies in poor countries. It is inappropriate to force the standard model on countries, but many poor countries will find it to be a useful prescription for economic progress as a means of addressing poverty, hunger, the environment, and inequities to women and other groups.

The standard model is the most reliable current prescription for economic progress but it is not perfect and needs continuing refinement. Businesses and commodity cycles have not been tamed. The IMF continues to be useful to respond to acute balance of payment crisis during the business cycle. The standard model depends on citizens in each country to use the growth dividend to provide public goods and a safety net for those who lack earning capacity (or lack transfers from others). Thus, the human element will continue to be critical in making development decisions.

## NOTES

1. John Williamson (1990) gave the label *Washington consensus* to the economic policy prescription being recommended by the World Bank, International Monetary Fund, and U.S. Department of Treasury for Latin America. At about the same time that Williamson proposed the Washington consensus, I was heading a task force to evaluate appropriate change in U.S. Agency for International Development (USAID) policies if the goal of the agency shifted from economic development to food security. The task force called for no major change in policy, but emphasized *broad-based* development policies that served both economic equity and efficiency. Also in contrast to the

Washington consensus, the *standard model* (originally called the *consensus synthesis* in the 1992 task force report (Tweeten et al. 1992)) relied on extensive economic literature and empirical evidence to prescribe policies that worked for any country but particularly for any developing country that wanted to pursue economic development with food security. I prefer the standard model to the consensus, but note with no surprise that the two models, though independently formulated, have many similarities. Perhaps this is one reason why Williamson in 1999 (see 2002, p. 4) stated that the consensus applies to Asia and Africa as well as to Latin America.

2. The IFPRI study (von Braun 1989, p. 4) cautions, nonetheless, that results vary from case to case. For example, commercialization of sugarcane production in the Philippines contributed to a landless class of former tenant corn farmers.
3. Unilateral liberalization refers to one country removing barriers to trade whether or not other countries reciprocate. Multilateral trade liberalization occurs when all countries reduce barriers through (say) negotiations of the World Trade Organization.

## REFERENCES

Bishop, M. "A Survey of International Finance." *Economist*, May 18, 2002, pp. 3-28.

Burfisher, M., ed. *The Road Ahead: Agricultural Policy Reform in the WTO–Summary Report.* Agricultural Economic Report No. 797. Washington, DC: Economic Research Service, USDA, January 2001.

Burfisher, M., S. Robinson, and K. Thierfelder. "Farm Policy Reform and Harmonization in the NAFTA." Pp. 66-74 in *Regional Trade Agreements and U.S. Agriculture.* Report No. 771. Edited by M. Burfisher, S. Robinson, and K. Thierfelder. Washington, DC: Economic Research Service, USDA, November 1998.

Crook, C. "A Survey of Globalization." *Economist*,September 29, 2001, pp. 3-29.

Diamond, J. *Guns, Germs, and Steel.* New York: W.W. Norton, 1997.

Diao, X., A. Somwaru, and T. Roe. "A Global Analysis of Agricultural Trade Reform in WTO Member Countries." *Background for Agricultural Trade Reform in the WTO: The Road Ahead.* ERS-E01-001. Washington, DC: Economic Research Service, USDA, 2001.

Diaz-Bonilla, E. and S. Robinson. "Shaping Globalization for Poverty Alleviation." *2020 Vision.* Focus 8, Policy Brief 1. Washington, DC: International Food Policy Research Institute, August 2001.

Epplin, F., and J. Musah. "A Representative Farm Planning Model for Liberia." Pp. 18-33 in *Proceedings of Liberian Agricultural Policy Seminar 1985.* Report B-23. Stillwater: Agricultural Policy Analysis Project, Department of Agricultural Economics, Oklahoma State University, 1987.

FAO. *Data of Food and Agriculture 1996.* Rome: Food and Agriculture Organization of the United Nations, 1996.

Friedman, T. *The Lexus and the Olive Tree.* New York: Anchor Books, 2000.

Gehlhar, M. "Multilateral and Regional Trade Reforms: A Global Assessment from a U.S. Perspective." In *Regional Trade Agreements and U.S. Agriculture,* edited by M. Burfisher and E. Jones . Agriculture Economic Report No. 771. Washington, DC: Economic Research Service, USDA, November 1998.

Gibson, P., J. Wainio, D. Whitley, and M. Bohman. *Profiles of Tariffs in Foreign Agricultural Markets.* AER No. 796. Washington, DC: Economic Research Service, USDA, 2001.

Hazel, P. "Shaping Globalization for Poverty Alleviation and Food Security: Technological Change: Introduction." *2020 Vision.* Focus 8, Policy Brief 8. Washington, DC: International Food Policy Research Institute, August 2001.

"How Much Should Petrol Be Taxed?" *Economist,* May 19, 2001, p. 69.

"Is It at Risk?" *Economist.* February 2, 2002, pp. 65-68.

Kanbur, R. "Shaping Globalization for Poverty Alleviation and Food Security: The Nature of Disagreements." *2020 Vision.* Focus 8, Policy Brief 2. Washington, DC: International Food Policy Research Institute, August 2001.

Makki, S., L. Tweeten, and J. Gleckler. "Agricultural Trade Negotiations as a Strategic Game." *Agricultural Economics* 10 (January 1994): 71-80.

Ratnesar, R. "Chaos Incorporated." *Time,* July 23, 2001, pp. 33-36.

Sumner, D. "Food Security, Trade, and Agricultural Commodity Policy." In *Agricultural Policy for the 21st Century*, edited by L. Tweeten and S. Thompson. Ames: Iowa State Press, 2002.

Tweeten, L. *Agricultural Trade*. Boulder, CO: Westview Press, 1992.

Tweeten, L. "Dodging a Malthusian Bullet in the 21st Century." *Agribusiness* 14, 1 (January/February 1998): 15-32.

Tweeten, L. "The Economics of Global Food Security." *Review of Agricultural Economics* 212 (Fall/Winter 1999): 473-88.

Tweeten, L. "Directions of U.S. Farm Policy under a Freer Trade Environment." (Presented at conference *The 2002 Farm Bill: Issues and Alternatives,* Fargo, ND, October 29, 2001.) Columbus: Department of Agricultural, Environmental, and Development Economics, Ohio State University, 2001.

Tweeten, L., and D. McClelland, eds. *Promoting Third-World Agricultural Development and Food Security*. Westport, CT: Praeger Publishers, 1997.

Tweeten, L., J. Mellor, S. Reutlinger, and J. Pines. *Food Security Discussion Paper.* PN-ABK-883. Washington, DC: Agency for International Development, 1992.

Tweeten, L., J. Sharples, and L. Evers-Smith. *Impact of CFTA/NAFTA on U.S. and Canadian Agriculture.* Working Paper 97-3. St. Paul, MN: International Agricultural Trade Research Consortium, March 1997.

U.S. Department of Agriculture. *Agricultural Outlook.* Washington, DC: Economic Research Service, USDA, June/July 2001, October 2001.

von Braun, J. "Commentary: Commercialization of Smallholder Agriculture." *IFPRI Report*, vol. 11, no. 2. Washington, DC: International Food Policy Research Institute, April 1989.

Wainio, J., P. Gibson, and D. Whitley. "Options for Reducing Agricultural Tariffs." *Background for Agricultural Reform in the WTO: The Road Ahead.* ERS-E01-001. Washington, DC: Economic Research Service, USDA, 2001.

Williamson, J. "What Washington Means by Policy Reform." In *Latin American Adjustment: How Much Has Happened?,* edited by J. Williamson . Washington, DC: Institute for International Economics, 1990.

Williamson, J. "What Should the Bank Think about the Washington Consensus?" Background paper prepared for the World Bank's *World Development Report 2000.* http://www.iie.com/williamson0799.htm, 2002.

World Bank. *World Development Report 1986.* New York: Oxford University Press, 1986.

"The World Social Forum." *Economist,* February 9, 2002, pp. 32-34.

# 4

# Radical Environmentalists

## INTRODUCTION

This chapter is about radical environmentalists, those who serially overstate and overdramatize environmental threats to food and agriculture. Such tactics are designed to and often do sway public opinion and public policy. Radical environmentalists also include zealots who, perhaps energized by overblown rhetoric, terrorize people and destroy property to "save the planet."

In this chapter, I briefly review ethics of the radical environmental movement. I then turn to trends in global food supply and demand, the impact of economic growth on the environment, and specific environmental problems affecting agriculture. The purpose is to discern if the doomsday rhetoric and violence of radical environmentalists are justified.

## ETHICS, ECONOMICS, AND THE ENVIRONMENT

The clash between radical environmentalists and conventional agriculturalists over environmental issues is as much over ethics as over facts. Radical environmentalists often identify with the postmodern morality outlined in chapter 2, placing protection of the environment over commandments of the Decalog.

Postmodern morality is apparent from a 1989 statement by "green" scientist Stephen Schneider in *Discover* ("Defending Science" 2002, p. 16):

> [We] are not just scientists but human beings as well. And like most people we'd like to see the world a better place...to do what we need to get some broad-based support, to capture the public's imagination. That, of course, means getting loads of media coverage. So we have to offer up scary scenarios, make simplified, dramatic statements, and make little mention of any doubts that we have. ...Each of us has to decide what the right balance is between being effective and being honest.

Thus, with what *The Economist* called "insufferable arrogance," Schneider and other "green" scientists make truth subordinate to saving the planet ("Defending Science" 2002, p. 16). Economics often gets shortchanged.

**Box 4.1**

It has been shown that the market maximizes output of an economy by encouraging use of resources to the point where incremental firm costs align with incremental returns. If social costs (benefits) differ from private costs (benefits), however, the market bringing the greatest net private benefits will not produce the greatest benefits to society. The *first law of welfare economics* is that public intervention (taxes, subsidies, regulations) in the market is warranted to align private with social incentives and thereby raise real output of society. The *second law of welfare economics* is that costs of public intervention must be less than the cost of market imperfections those interventions were intended to correct. Many public interventions to correct market flaws for the sake of the environment fail that second law.

The difference between private and social cost (benefit) is called an *externality*. In the case of environmental services, the market is likely to register *use value* (satisfaction from directly consuming a service), but is less likely to register *existence value* (e.g., the satisfaction of preserving a scenic view or species that the individual never will see) or *option value* (preserving a service not used today but may be used in the future). Markets work well for use value, but work less well to allocate resources serving existence and option values. The latter are externalities. The *travel-cost model* and *contingent valuation model* are but two of the many frameworks developed by economists to value nonmarket services and thereby incorporate externalities into benefit-cost analysis for making better decisions regarding allocation of environmental services.

Existence and option values are related to the concept of *safe minimum standard* and the *precautionary* principle. The safe minimum standard concept recognizes, for example, that some threshold number of individuals must be preserved for a species to avoid extinction. The precautionary principle holds that "it is better to be safe than sorry" so that a mistake of species overprotection is preferred to a mistake of underprotection.

Radical environmentalists are highly critical of the free enterprise system "producing for profit rather than for people." But production for profit serves people if environmental externalities are identified and corrected. While the competitive market turns profit seeking into public good that was no part of the firm's intention, no such invisible hand guides nonmarket agents (including environmentalists?) to serve society. In the case of *open access property* such as oceans, tropical forests, and the earth's atmosphere, individuals and firms pursuing their self-interest are collectively made worse off in what is called the *prisoners' dilemma.*

**Economics and the Environment**
Markets alone do not properly allocate environmental resources. A public role and hence politics are critical (see Box 4.1 for conceptual framework). Knowing this and bent on achieving their objectives in the absence of reliable data, deep environmentalists cannot resist spreading misinformation in the media. That effort generates public anxiety regarding the state of the planet. Emotionalism motivates people to fill coffers of environment organizations, but it does not generate sound environmental policy.

Radical environmentalists and economists think differently about ecology. To be sure, greens and economists alike recognize the intrinsic value of pristine air, water, soil, wildlife, and biodiversity. Radical environmentalists refer to our ecological heritage as "priceless" and as "too important to compromise at any cost." Economists also recognize the cost–a pristine world is expensive in other, forgone goods and services. Society cannot accomplish everything. Priorities must be set. In doing so, it is useful to weigh incremental costs versus benefits of interventions to protect nature. The cost of having no soil erosion, no net carbon release into the atmosphere, and no chemicals in the water supply is infinite. As greens themselves acknowledge, the earth kept in a somewhat pristine state has carrying capacity for only a fraction of today's population.

Carrying capacity cannot be separated from the issue of sustainability, where sustainable is defined in the UN *Brundtland Report* (WCED 1987) as development that "meets the needs of the present without compromising the ability of future generations to meet their own needs". Most of us are not satisfied to see future generations live only as well as we; we are determined to see that they will live even better! Thinking mere "sustainability" is thinking too small!

But how can they live better if we are using up the earth's resources? The answer as noted by Nobel Laureate Robert Solow (1986) is that we should leave future generations with human and technological capital so that they can enjoy a better quality of life. An altogether attainable goal is to use today's resources to create human, technological, and social capital so that future generations can enjoy a higher standard of living by getting more out of fewer natural resources.

The remainder of this chapter explores whether this planet and, more specifically, U.S. and world agriculture and food production are sustainable. This chapter reviews progress in addressing environmental problems of agriculture. One purpose of this exercise is to judge if the claims of radical environmentalists are justified. Before we examine the global food supply-demand balance, however, the reader is encouraged to digress by reading Box 4.2 regarding *The Skeptical Environmentalist* (Lomborg 2001), a controversial book closely related to issues addressed in the remainder of this chapter.

## FUTURE GLOBAL FOOD SUPPLY-DEMAND BALANCE
Before looking at actual and prospective trends in food supply and demand, we note the gloomy global prospects for food portrayed by deep environmentalists. Pessimism regarding ability of the world to feed itself goes back at least to Thomas Malthus who in 1798 published *An Essay on the Principle of Population*. He

**Box 4.2**

*The Skeptical Environmentalist* is of unprecedented scope and documentation. In it Bjorn Lomborg (2001), a Danish statistician, critically reviews the scientific and popular literature on major environmental issues. Some of the findings of Lomborg's scholarship are cited in this chapter because of the extensive research review upon which he draws his conclusions.

As a statistician and environmental generalist, Lomborg has the advantage of breadth, although not of the depth in subdisciplines possessed by scientific specialists. The scholarly generalist approach enabled him to uncover a recurring pattern in the environmental literature: a tendency of a vocal few radical environmental agencies and individuals to dominate and "spin" the interpretation of (often) objective findings from environmental science.

Radical environmentalists have adroitly applied the maxim "bad news drives out good news." They have consistently overrated threats to the environment and understated the world's progress in protecting the environment. The problem is not with the vast majority of dedicated scientists who do yeoman service in identifying environmental problems and options to address them, but with the few but outspoken radical organizations and individuals who through malice, misplaced idealism, or wanton ignorance mislead the public.

As expected, the green community mounted a massive assault to discredit the work of Lomborg. Kathryn Schultz (December 2001), assistant editor of *Grist*, an environmental magazine that devoted a series of articles to attacking *The Skeptical Environmentalist,* accused Lomborg of splitting the left by his claim that billions of dollars ineffectively spent to stop global warming would come partly at the expense of high-payoff education and health spending on the disadvantaged. E. O. Wilson (December 2001) contended that the extinction rate is ten times Lomborg's estimate of 0.014 percent per year. Norman Myers (December 2001) faulted Lomborg for "utterly failing to acknowledge it [extinction] as a phenomenon that will impoverish the planet for millions of years to come." Stephen Schneider (December 2001), perhaps unhappy that Lomborg had criticized his work, petulantly lamented "What a monumental waste of busy people's time countering the scores upon scores of straw men, misquotes, unbalanced statements, and selective inattention to the full literature."

Given the comprehensiveness of Lomborg's work, one can be impressed with how few errors were turned up by critics and how much they had to rely on personal attacks. One of the most frequent complaints was that Lomborg is an academic statistician rather than an environmental scientist. One can argue, however, that the frequently inadequate and conflicting nature of data on the environment makes statistics a useful discipline for someone attempting to sort fact from fiction in the environmental literature.

contended that population grows geometrically (1, 2, 4, 8, 16, etc.) whereas food production grows arithmetically (1, 2, 3, 4, 5, etc.). Left unchecked by pestilence, wars, or other interventions, population inevitably will outrun the food supply. Radical environmentalists have persisted in that theme to this day.
Paddock and Paddock (1967) wrote:

> The famines which are now approaching will not, in contrast, be caused by weather variations and therefore will not be ended in a year or so by the return of normal rainfall. They will last for years, perhaps decades, and they are, for surety, inevitable. ...In fifteen years the famines will be catastrophic and revolutions and social turmoil and economic upheavals will sweep across Asia, Africa, and Latin America. (p. 8)

Paul Ehrlich (1968) in his prologue to *The Population Bomb* concluded that

> the battle to feed all humanity is over. In the 1970s the world will undergo famines–hundreds of millions of people are going to starve to death in spite of any crash program embarked upon now. At this date nothing can prevent a substantial increase in the world death rate.

Lester Brown, long head of the Worldwatch Institute and senior editor of *State of the World,* has been a consistent Malthusian since he worked for Secretary of Agriculture Orville Freeman during the "world food crisis" of 1966-67. In 1974 he predicted "that this era [of food abundance] is ending and is being replaced by a period of more or less constant scarcity and higher prices", a claim he again made in 1996 (cf. Lomborg 2001, p. 108). In 2000 the Worldwatch Institute (WI 2000, p. xvii) proclaimed that

> we are about to enter a new century having solved few of these [environmental] problems, and facing even more profound challenges to the future of the global economy. The bright promise of a new millennium is now clouded by unprecedented threats to humanity's future.

Worldwatch Institute contended that as "As the global economy expands, local ecosystems are collapsing at an accelerating pace" in forms such as water and food shortages, disease, and ethnic and political conflicts (WI 2000, p. 4). As we shall observe in the following pages, economic expansion is a cure for rather than the cause of some of the most severe world food and environmental problems.

Al Gore (1992, pp. 269, 273) seemed to take a page from Brown when he wrote in *Earth in the Balance* that "modern industrial civilization as presently organized is colliding violently with our planet's ecological system." He goes on to warn of "a steady stream of progressively more serious ecological catastrophes that will be repeatedly proffered to us."

None of the foregoing jeremiads came close to reality: the world has been in fact better fed in recent decades than ever before. If modern agriculture is unsustainable and the world will be unable to feed itself, then environmentalists might have at least some justification for strong and deceptive rhetoric to awaken the masses to the sacrifices and policy interventions required to save themselves.

Before examining specific resource limitations and environmental problems, in this section I review findings of a study I published tracing historic trends and projecting long-term global food supply and demand (Tweeten 1998). Principal conclusions of that study are:

- Percentage rates of increase in global yields of major crop and livestock groups are falling. Yield trends since the 1950s as recorded by the Food and Agricultural Organization (FAO 2000) of the United Nations have fluctuated around a remarkably linear (straight-line) trend. Figure 4.1 shows the yield trend for cereals. Cereals account for over half of all calories, thus it is not surprising that the percentage annual increases depicted along the horizontal axis are nearly the same for cereals and an aggregate of all foods. Percentage annual trend yield increments fell from 3.13 in 1961 to 1.92 in 1981 to 1.43 in 1999 (fig. 4.1).
- Global land in crops, after increasing approximately 1 percent annually for several decades, has remained quite stable for a decade. That is, gains from new cropland in Brazil and elsewhere have been offset by loss of cropland from urbanization, environmental degradation, and political change (e.g., demise of the Soviet Union). With constant land area and slowing yield growth, global food production could grow slower than population.
- A felicitous development is that this Malthusian specter of rising real food prices (of special concern to poor countries) is likely to be narrowly averted by a profound shift in global demographics. All major demographic projections look to a continuing slowing of global population growth until population begins to decline within the century.

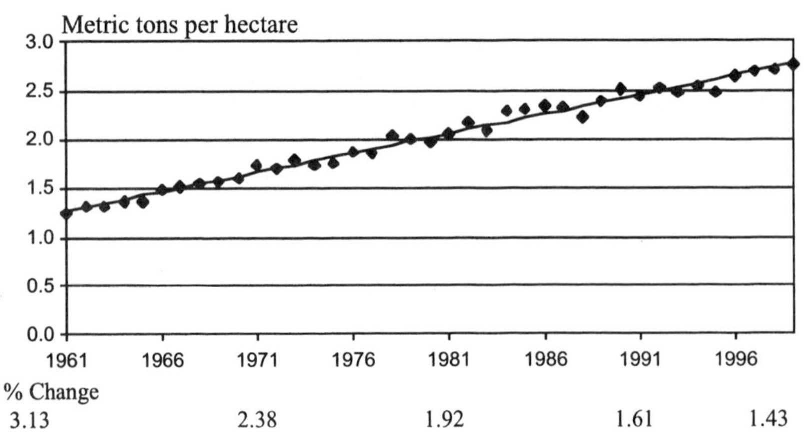


**Fig. 4.1. World Cereal Yield Trend From 1961 To 1999 (From FAO 2000).**

Since the 1950s, the total fertility rate (TFR, the expected number of children a woman would bear throughout her life) has fallen in every region of the world. From an average of near 6 children per woman in 1950-55, TFR by 1990-95 fell to 3.4 in India, 3.5 in "other Asia," and to 3.0 in Latin America (UN 1998, p. 11). Rates continue to fall (UN, p. 8). TFRs in the 1990-95 period were below the 2.1 average children per woman needed to sustain population over the long run in Europe (1.57), China (1.92), and North America (United States and Canada) (2.02).

The medium United Nations (UN) population projection is a widely used demographic forecast, but appears to unrealistically assume that TFRs will converge to 2.1 in both developed and developing countries. That assumption overestimates future population according to Lutz et al. (1996, p. 365):

> The United Nations and other institutions preparing population forecasts assumed that fertility would increase to replacement level and that subreplacement fertility was only a transitory phenomenon. ...It is difficult, however, to find many researchers who support this view. Too much evidence points toward low fertility. Many significant arguments support an assumption of further declining fertility levels. They range from the weakening of the family in terms of both declining marriage rates and high divorce rates, to the increasing independence and career orientation of women, and to a value change toward materialism and consumerism. These factors, together with increasing demands and personal expectations for attention, time, and also money to be given to children, are likely to result in fewer couples having more than one or two children and an increasing number of childless women. Also, the proportion of unplanned pregnancies is still high, and future improvements in contraceptive methods are possible. The bulk of evidence suggests that fertility will remain low or further decline in today's industrialized societies.

The UN's low/medium scenario seems more realistic. It presumes continuation of TFR trends, but converging to 1.9 TFR for all regions by year 2025. This scenario projects a peak world population of 8.0 billion people in 2050, declining to 6.4 billion by 2150 (UN 1998, p. 14).

Several peak population projections are summarized as follows:

| Source | Peak Global Numbers (billions) | Population Year |
|---|---|---|
| World Bank (Bos et al. 1994) | 11.3 | 2128 |
| International Institute for Applied Systems Analysis (Lutz et al. 1996, p. 376) | 10.8 | 2080 |
| UN (low/medium scenario, 1998) | 8.0 | 2050 |

Most projections point to peak global population in less than a century, followed by population decline. Today's population is not expected to double before global population peaks.

The medium UN projection may overestimate future TFRs (2.1) in developed countries, but the low/medium UN projection may underestimate future TFRs (1.9) in developing countries. Hence, both UN scenarios are employed in projecting food demand in figure 4.2.

Future portents for food consumers would be ominous indeed in the absence of falling fertility rates. Figure 4.2 shows projected aggregate food supply based on a continuation of 1961-99 yield trends and no net increase in global cropland. Alternative aggregate food demand projections from 2000 to 2150 are from the indicated population projections coupled with a 0.3 percent annual increase in food demand per capita due to income growth.

If population and income would maintain their 1995 to 2000 trend growth, future demand would sharply outgrow future food supply (fig. 4.2). Real commodity prices would need to rise to draw additional land and other resources into food production and to restrain consumption. If instead the United Nations *medium* population projection demand scenario or the IIASA (Lutz et al. 1996) scenario prevails, food demand growth is projected to modestly outstrip food supply growth until approximately 2075, a gap that could probably be covered by small increases in real food prices.

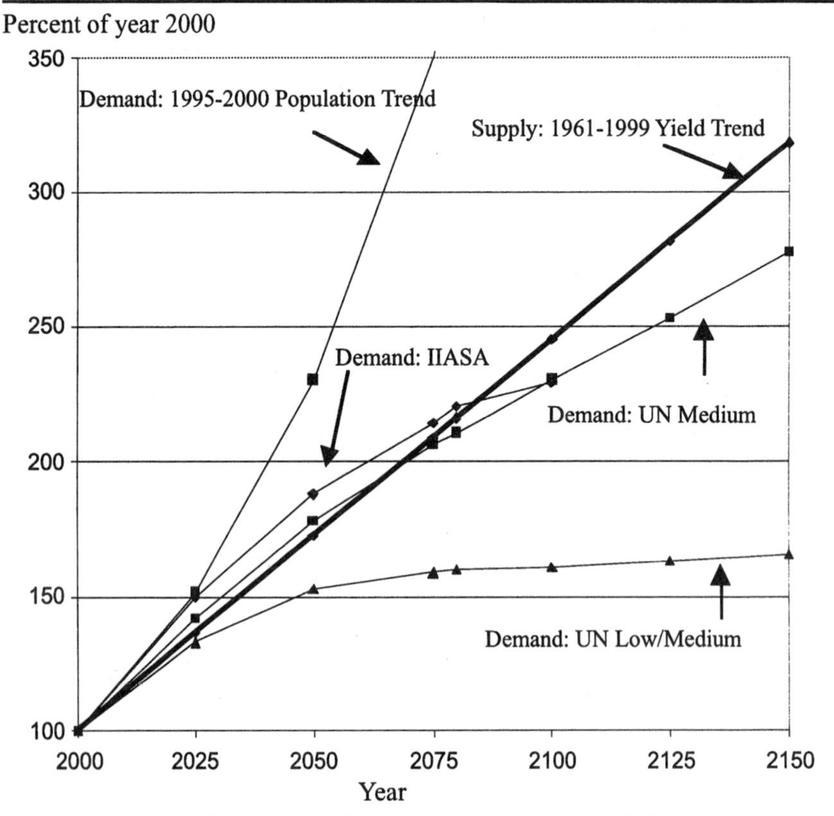


**Fig. 4.2. Projected Global Food Supply And Demand Trends Under Alternative Scenarios, 2000-2150 (from Tweeten 1998 and text).**

If the demand scenario is based on the perhaps more realistic United Nations *low/medium* population projection, food supply grows faster than food demand. This outcome would allow real farm commodity prices to continue to fall and perhaps even accelerate the rate of decline (fig. 4.2). It follows that slowing population growth may narrowly avert rising real farm commodity and food prices in the first half of the twenty-first century.

The UN population projections of figure 4.2 were made in 1998 and revised in 2000. Based on falling fertility rates, the UN lowered its median world population projection for 2050 from 9.4 billion down to 9.3 billion persons (UN 2002). The 1998 UN projections were used to construct figure 4.2 because they extend farther into the future than the 2000 projections, but the revised projections reinforce the conclusion that global food supply is likely to outpace food demand growth. Real food prices at the farm level seem destined to continue their secular decline. Nonetheless, the food balance could be so tight that the world will welcome emerging technologies such as genetic engineering.

## ECONOMIC PROGRESS AND THE ENVIRONMENT

The conclusion from the above analysis is that long-term global food supply-demand prospects call neither for panic nor complacency. Food will be available to countries and individuals with purchasing power (income). At issue is whether rising income is compatible with protecting the environment as well as food security.

A study by Hervani and Tweeten (2002) indicates that fortuitous interaction between declining population growth and high income ultimately saves the environment. (Effective policies to establish appropriate incentives and institutions must continue to accompany affluence.) Lower birth and population growth rates, brought about in part by higher income, eventually reduce pressure on the environment. Higher income provides savings out of current income no longer needed to provide necessities. The savings in turn can be used to finance investment in science and technology that reduces pressure on the environment. Rapid productivity gains of agriculture made feasible by agricultural research financed out of economic growth enables cropping of fewer and environmentally safer acres while freeing cropland for grass, trees, recreation, and biodiversity.

Education and research made possible by economic progress promotes awareness of environmental hazards, which in turn can generate effective policy responses. Higher-income consumers demand greater efforts to protect the environment because, once their basic needs are met, they manifest a high income elasticity for environmental quality. Furthermore, as their incomes rise, consumers spend a larger share on *services*, whose production and disposal are less detrimental to the environment than are the production and disposal of *goods*. The result is a trajectory of increasingly less environmental damage and greater environmental preservation per unit of national income and per person as per capita income rises.

The relationship between per capita income and environmental degradation has been conceptualized as an environmental *Kuznets Curve* (Seldon and Song 1994). Empirical studies support an inverted U-shaped relationship, that is, environmental degradation first rises and then falls as per capita income rises under economic growth.

In a comprehensive empirical study of the Kuznets Curve, Hervani and Tweeten (2002, ch. 11) regressed environmental and natural resource variables on income per capita, population density, and selected other variables for approximately 120 countries using data from the 1990s. To illustrate the results, consider the most abundant greenhouse gas (GHG), carbon dioxide ($CO_2$). From a base of $8,000 per capita income, a 1 percent increase in per capita income raises $CO_2$ emissions per capita by 2.66 percent. This elasticity declines to 1.93 percent at an income of $16,000 per capita and it becomes negative at $20,000 per capita. Thus, emissions per capita fall at higher income levels.

Other environmental and natural resource variables also show the anticipated pattern with respect to per capita income. Emissions per capita reach a peak (rate of growth is zero) at approximately $16,000 per capita for nitrogen dioxide ($NO_2$) and sulfur dioxide ($SO_2$), at approximately $24,000 per capita income for methane, and at approximately $2,000 per capita for suspended particles such as carbon soot (Hervani and Tweeten 2002, pp. 224-28).

Per capita use of energy and oil increases up to approximately $18,000 per capita. Use of phosphate fertilizer per capita begins to decrease at approximately $14,000 per capita. Organic components in water are both an indication of water resource degradation and of natural resource depletion, the latter because resources are depleted to produce fertilizers and other chemicals found in water. Fertilizers in turn raise levels of algae and other organisms in water. The income elasticity implies that added income above approximately $20,000 per capita reduces annual organic water pollution per capita in the long run.

The above numbers were for per capita emissions. Total emissions for a country are per capita emissions multiplied by population. To illustrate for $CO_2$, at $16,000 per capita, adding the elasticity of population growth with respect to income (-1.21) to the elasticity of $CO_2$ with respect to income (1.93) indicates that a 1 percent rise in income at $16,000 raises *total* national $CO_2$ emissions by 1.93-1.21 or only 0.72 percent. Including the impact of income on population brings the turning point to a better environment at a lower income level. Thus, higher per capita income and slower population growth work together synergistically to hasten the move not only to lower *per capita* emissions and use of natural resources, but also to lower *total* emissions and natural resource depletion.

In summary, when the impact of income on population growth is considered, rising income per capita beyond approximately $15,000 improves most of the nine environment variables considered by Hervani and Tweeten. Particles, one of the most unhealthy air pollutants, begin to go down at much lower income levels. Results are consistent with those from numerous studies reviewed by Hervani and Tweeten (2002).

Income per capita in numerous industrial nations exceeds that at which pollution turns around, but the conclusion that environmental protection predominates only at quite high per capita income is troublesome. However, the technologies and institutions developed in wealthy counties also help poor counties. Especially notable is that particle pollution declined over 20 percent for countries over most *all* per capita income levels between 1972 and 1986 (Shafik 1994, p. 764; World Bank 1992, p. 41).

Still, countries in Africa and South Asia cannot easily achieve the per capita income levels that protect the environment. Much resource degradation and depletion occurs as large countries such as India and China attempt to pass through the environmental transition. It follows that special efforts at education, technology transfer, and proper policy are required for poor countries to make the environmental transition with as little injury to the environment as possible. Such an effort requires public resources that in turn require a tax base of income.

Economic progress has other advantages to protect the environment As noted above, affluence generated by economic growth promises to stabilize global population at approximately 11 billion people in the twenty-first century. This reduces pressure on agricultural resources. Countries following standard economic model policies (see chapter 3) attract foreign investment that promotes economic growth. The wealth generated thereby makes more funds available for protecting the environment or income transfers to the poor.

Nations that have made economic progress do not have hungry peasants desperate for new cropland to assuage their hunger. Such nations can hire enough civil servants to help preserve sensitive ecosystems and biodiversity. Better-paid civil servants and government officials minimize corruption that brings logging in rain forests, opening the land for hungry settlers who convert forestland to cropland. Countries making economic progress do not need to engage in an international "race to the bottom," competing for multinational corporation jobs by offering lax environmental rules.

## SPECIFIC IMPEDIMENTS TO AGRICULTURAL SUSTAINABILITY

Factors potentially thwarting agriculture's sustainability include global warming, loss of biodiversity, and shortages of water, land, energy, and phosphate. Some environmentalists and laypersons also view contamination of land, air, water, and food by agricultural pesticides and other chemicals as a serious threat to sustainable food and fiber production. This section reviews the case for each of these concerns. This review concludes that none of these concerns need interfere with sustaining an ever more productive and environmentally sound agriculture in the twenty-first century and beyond.

### Global Warming

Global warming is perhaps the single most vexing international environmental issue of our time. Evidence is unequivocal that greenhouse gases such as carbon dioxide, nitrous oxide, and methane are accumulating in greater concentration in the atmosphere. Evidence is compelling that that accumulation has some anthropogenic (manmade) origins, especially deriving from combustion of fossil fuels. Surface temperature increased approximately 0.7°C in the past century, and is projected to increase 1 to 6°C in the next century if current policies continue (IPCC 2001). Predictions are imprecise for many reasons, but especially because of as yet unpredictable interactions among greenhouse gases, cloud formation, ocean life and currents, and particles such as sulfur aerosols.

Emerging technologies could reduce the actual temperature increment to less than the predicted high estimate (6°C). Fossil fuel reserves are adequate to power the world economy for hundreds of years, but technological advances in "clean" wind, solar, and nuclear energy sources seem likely to be economically competitive with and hence will substitute for fossil fuels long before fossil fuel reserves are exhausted. Carbon dioxide release into the atmosphere will fall as fossil fuel consumption drops.

Solar power cost per unit has been declining at a rate of 50 percent per decade. If solar power cost declines 30 percent per decade, "clean" energy may replace fossil fuels by approximately 2100 (Chakravorty, Roumasset, and Tse 1997, pp. 1222, 1223). The replacement would take place earlier with the same 30 percent per decade decline in solar energy cost if a tax of $100 per ton is placed on carbon emissions. The carbon tax not only would diminish global warming, but also would reduce traffic congestion, urban sprawl, cropland loss, and dependence on imported oil.

The cost of unmitigated global warming to developed countries such as the United States is estimated to be 1-1.5 percent of GDP (IPCC 1996, pp. 183-89). If implemented and sustained, the Kyoto Protocol could reduce global temperature only about 0.15°C by 2100 (Parry et al. 1998, p. 286), and could reduce the rise in the sea level by 2.5 centimeters or 1 inch (Wigley 1998) *below the rise with no protocol*. Thus, the protocol would cut the projected sea level rise of 25 centimeters by only one-tenth.

The modest benefit from the Kyoto Protocol arises in part because developing countries are not included. The Kyoto Protocol would require a reduction of U.S. greenhouse gas emissions by 7 percent from their 1990 level by 2008-12. For fossil fuels, that reduction might translate into more than a 30 percent drop below business-as-usual levels by year 2012. The annual cost to the nation could be $100-$200 billion–1 to 2 percent of national income (Lomborg 2001, p. 287). Thus, the cost to the United States of global warming could be similar with or without the protocol. Supplementing the current U.S. research and development outlays of $200 million per year on clean energy with revenues from a tax on fossil fuels might be a way to reach Kyoto greenhouse gas targets at less cost to the public.

Global warming will have varied impacts on agriculture. Poor countries at or near the tropics fare worst although temperatures are expected to rise more in regions distant from the tropics. Atmospheric carbon dioxide acts like a fertilizer to plants. With moderate adaptation of varieties and cultural practices by farmers, global warming of a 2.5 to 5.2°C magnitude is projected to reduce cereal output in developing countries 6-7 percent and raise cereal output in developed countries 4-14 percent–for virtually no net change in cereal production for the world as a whole (Rosenzweig and Perry 1994, p. 136; see also Mendelsohn and Neumann 1999).

Models used to project global warming have difficulty accounting for the expected greater temperature increases in colder than in warmer climates, in nighttime more than daytime, and in winter more than summer. Food production could expand in Canada and Russia while it might fall in Nigeria and Indonesia. Tropical countries and lowlands such as in Bangladesh will be adversely affected, and many people will need to be relocated to higher ground.

In conclusion, global warming is of concern, but it does not justify panic. Appropriate policy responses include carbon emission taxes and/or formation of "cap and trade" emission markets. The latter would allocate reduced emission allowances to firms and individuals based on history or bids. Trading in emissions rights would follow, presumably with firms having high-value carbon uses purchasing emission rights from firms with low-value carbon use. And firms with low cost of cutting emissions would sell emission permits to firms with high cost of cutting emissions. Sulfur dioxide emissions markets have already operated with success, demonstrating that markets, in contrast to command and control policies, can be used to efficiently reduce emissions.

Payoffs could be large from research to reduce costs of solar, wind, nuclear, and other clean energy to replace fossil fuels. Bioengineering and other agricultural research could adapt crops to the stresses and opportunities offered by new climatic conditions.

Forestland, expanding particularly in North America and Europe, has potential to sequester carbon. One study by Tanner (2001) indicates that no-tillage farmland cropping systems have potential to profitably sequester carbon, but only modest amounts relative to forestland. Similarly, production of ethanol from corn or biodiesel fuel from soybeans has limitations because current technology requires nearly a gallon of petroleum (plus lots of other resources) to produce a gallon of biofuel.

**Biodiversity**

Loss of biodiversity ranks with global warming as a major environmental concern to American people. Humans have long been a source of extinction. Jared Diamond (1997) argues that Native Americans wiped out several large animal species that might have been domesticated to materially increase productivity of agriculture. (These animals had evolved before humans arrived, thus had no natural fear of hunters who found them easy prey.)

The seriousness of plant and animal extinction is difficult to appraise–scientists do not even know with reliability the number of existing species or extinct species. Estimated numbers of extant species range from 2 million to 80 million. Estimated extinctions range from 2 to 250,000 species per year (Lomborg 2001, pp. 249, 251). The latter estimate, from Paul Ehrlich, implied that all species would be gone in a few years. Ehrlich and Ehrlich (1996, pp. 112, 113) state that "biologists don't need to know how many species there are, how they are related to one another, or how many disappear annually to recognize that the earth's biota is entering a gigantic spasm of extinction." Such illogic does not help clarify the biodiversity debate.

Based on the work of several scientists, Lomborg (2001, p. 255) concluded that the best-guess extinction rate for animals is 0.7 percent each 50 years. This is not a trivial rate: it is 1,500 times the natural background extinction rate although it is far below the rates advanced by Ehrlich and others. As indicated earlier, E. O. Wilson (2001) is critical of Lomborg's estimate, contending that the extinction rate is ten times higher.

Whatever the current extinction rate, it may fall in the future. The major enemy of species preservation is poverty coupled with rapid population growth. Higher per capita income, falling birthrates, plantation forests, urbanization, and high-yield agriculture reduce pressure to expand cropland and log forests rich in biodiversity. Thus, economic progress above some income threshold probably reduces species loss.

Promising measures to preserve biodiversity include:

- Increasing agricultural productivity through research and development so that food demand can be met from production on environmentally safe land. Hence, cropland need not be extended to marginally economic tropical forestland and other biodiverse ecosystems.
- Increase productivity of forests, including tree plantations especially productive in transition zones between tropical rain forests troubled by insects and disease but rich in biodiversity, and more arid and artic lands troubled by slow tree growth. Producing in efficient tree plantations reduces prices of timber products so that harvesting rain forests and boreal forests becomes uneconomic.
- Set aside parks and preserves in biodiverse areas. Many of the richest areas ecologically are the poorest economically. Ecotourism offers promise to finance some preserves. Affluent nations can give technical and financial assistance to poor countries for preserving wildlife ecosystems.
- Support banks preserving plant and animal genetic materials. Modern preservation techniques can almost indefinitely store such materials for subsequent bioengineering research and development.

Drastic measures to preserve biodiversity are not called for. An example is the Buffalo Commons to restore the Great Plains to its "wild" state. Another example is the Wildlands Project proposed by environmentalists to move much of the United States' population to densely populated "city islands" while restoring natural wilderness in the vast hinterlands of the rest of the nation. That proposal would impose huge social costs relative to benefits.

Forming a reliable estimate of the value of preserving a species is about as elusive as estimating extinction rates. Species lack markets and hence lack prices to reflect value as a tradeoff between supply and demand. Furthermore, the appropriate measure of value is *marginal* rather than *average* worth employed in past attempts to value nature. Nature has major redundancy–chimpanzees and humans share over 98 percent of their genes. With modern bioengineering, gene transfer is possible *among* species. This differs from traditional breeding that transfers genes *within* species. Modern biotechnology raises the value of unique genes that have diverse applications, but also reduces the value of species that have a redundant gene structure.

The marginal value of biodiversity is often low because a plant or animal may provide little new genetic material of value and because searchers cannot afford to examine the genes of more than a few of the many extant species. Keystone species occupying critical junctures in ecosystems are key to survival of other species. These considerations suggest that preservation of keystone species with unique features is important, and that requires attention to ecosystem habitats.

**Soil Erosion**

Global warming and biodiversity loss may be the greatest global environmental concerns, but soil erosion historically has been the foremost environmental challenge facing American agriculture. Fortunately, much progress has been made in cutting erosion in the United States: the average sheet and rill (water) erosion rate has fallen from 23 tons per hectare in 1938 (Magleby et al. 1995; U.S. Department of Agriculture 1938, p. 595) to nearly 7 tons per hectare in the mid-1990s (Pierce and Nowak 1994, p. 2). Water erosion averaged 5.3 tons per hectare in 1997 and wind erosion averaged 6.5 tons per hectare, the latter down from 9.3 tons per hectare a decade earlier.

Crosson (1992, p. 196) estimated that continuation of erosion at current rates would reduce U.S. agricultural productivity by 3 to 10 percent in a century. Cropland fell 0.4 million hectares or 0.2 percent annually from 1982 to 1992 (U.S. Department of Agriculture 2000, p. IX-9). Thus, annual farm output could drop 5 percent from erosion and 20 percent from urban encroachment in a century, or on average by 0.25 percent per year. This could be offset eight times over by future productivity advances, if future productivity grows at the 1950-96 average of 2 percent per year (Council of Economic Advisors 1996, p. 388). This comparison highlights the nation's progress in soil conservation and the importance of productivity gains through science and technology. It does not abrogate the need to control losses of farmland, but heroic measures are unnecessary.

**Minerals and Energy**

The sustainability of agriculture depends on critical minerals and petroleum. Reserves of nitrogen, potassium (potash), and energy (from petroleum, tar sands, shale, coal, nuclear, wind, solar sources) are adequate for centuries. Greenhouse gas pollution is a problem with some of these sources as discussed elsewhere in this chapter.

The most limiting mineral for commercial fertilizers may be phosphorus (phosphate). It is an essential building block for life and has no substitutes. Paul Barton (1996) of the U.S. Geological survey reported global reserves of phosphorus (3,400 mmt) relative to 1992 use (14 mmt). The 1992 consumption continued for 242 years would exhaust reserves. Projections that account for rising annual consumption further shorten the projected life of reserves (Tweeten and Amponsah 1998).

Recent discoveries of phosphate off the coast of Florida have doubled estimates of phosphate reserves. The finding highlights an important principle: actual reserves are likely to be very much higher than known reserves of minerals whose prices have been low for extended periods. Shortages of phosphate would energize a search for reserves likely to meet with much success.

Metal scarcity is declining based on market trends apparent in real prices. In 1980, Julian Simon, an economist who has argued persuasively that material conditions of life have been improving for most people over most of the world, challenged radical environmentalist Paul Ehrlich and colleagues at Stanford University to a bet that the real price of metals would fall in the 1980s (Simon 1996, pp. 35, 36). Not only did the index of metal prices fall so that Simon won the $10,000 bet, the price of each metal in the index fell (see Box 4.3).

**Box 4.3**

This vignette suggests an opportunity for levity while discouraging serial bias in the forecasts by prophets of doom. My proposal is that environmentalists and economists subject their forecasts to all who would like to bet against them. Serial mendacity or incompetence would be punished by depletion of ones bank account. A depleted bank account would chasten those who exploit a public seeking bad news about the environment, and would give an economic incentive to correct bias.

**Chemical Contamination of Food and Water**

Dangers of chemical contamination of food and water often have been overstated. Rachel Carson, the mother of modern environmentalism, started things off in *Silent Spring* with her interpretation of a once idyllic town in the heart of America. The community was her mental construct or composite of the things that she claimed had actually happened in one or more communities after agricultural pesticides were introduced. She wrote (1962, p. 2):

> Then a strange blight crept over the area and everything began to change. Some evil spell had settled on the community: mysterious maladies swept the flocks of chickens; the cattle and sheep sickened and died. Everywhere was the shadow of death. The farmers spoke of much illness among their families. In town the doctors had become more and more puzzled by new kinds of sickness appearing among their patients. There had been several sudden and unexplained deaths, not only among adults, but even among children, who would be stricken suddenly while at play and die within a few hours.
>
> There was a strange stillness. The birds, for example–where had they gone?...The few birds seen anywhere were moribund; they trembled violently and could not fly. It was a spring without voices.

Such prose changed policy. No one knows whether DDT would have been banned and pesticide regulations strengthened without such rhetoric. At any rate, it is useful to look at how well Carson's description applies to today's America.

Based on a 1990 survey (EPA 1990), approximately 2 percent of water wells had nitrogen and 0.6 percent of wells have synthetic pesticides above levels judged safe by the Environmental Protection Agency. While the nation would like to have no chemicals above safe levels in wells, it is difficult to find any case of morbidity or mortality due to such contamination.

In early 2002, the U.S. Geological Survey reported results of measurement for 95 possible contaminants in the nation's streams. With today's highly precise instruments to measure chemicals, it is not surprising that the survey found almost all the contaminants it was looking for. In addition to farm and urban-origin pesticides, it found household-source contaminants including antibiotics, steroids, caffeine, antibacterial agents, and insect repellants. Such findings provide much grist for radical environmentalist mills. Many people will be driven by their rhetoric to alternative drinking water sources. Several points need to be considered, however:

- The minute levels of chemicals found except in rare cases pose no threat to human health. The ancient maxim applies: "The dose makes the poison."
- Any contaminants in stream waters rarely reach drinking water. Water treatment plants monitor contaminant levels and provide effective treatment to maintain safety.
- Public regulatory agencies monitor and upgrade if necessary safe minimum standards for contaminants. Since 1942, federal regulations allowed up to 50 parts of arsenic per billion parts of water in municipal water systems. The Environmental Protection Agency will reduce the maximum allowable level to 10 parts per billion by year 2006 (Stratton 2002, p. 12).

Americans spent $2 billion on bottled water and a like amount on home water treatment devices in 1998 (Stratton 2002, p. 12). Motivated in part by rhetoric from radical environmentalists, that market is growing rapidly. Yet tap water is better for many if not most people because it contains fluoride effective in reducing tooth decay.

The United States has been so successful in controlling synthetic chemical contamination of food supplies that American consumers get over 10,000 times as much carcinogens from natural as from synthetic sources (Ames and Gold 1989)! And even the natural carcinogens in food are not considered to be a problem. Age-adjusted cancer deaths are falling for gastrointestinal and other cancers on average with the exception of lung and other cancers attributed to smoking. Pesticides are inconsequential sources of cancer deaths based on studies cited by Lomborg (2001, pp. 232-36).

Nearly half of all chemicals in foods are rodent carcinogenic–they cause cancer in rodents ingesting massive quantities over a considerable period of time. To protect themselves from predators, plants, unlike animals unable to flee predators, are especially high in natural toxins. There are 1,000 natural chemicals in coffee, for example. Fruits and vegetables have high levels of toxins but also contain generous amounts of antioxidants that protect people from cancer. Experts such as Ames and Gold (1989) note that healthy eating calls for more, not less, consumption of fruits and vegetables.

Radical environmentalists can raise cancer rates by planting unwarranted fear in consumers of chemicals in fruits and vegetables. That fear can discourage consumption directly. Or it can cut fruit and vegetable consumption because of higher prices generated by the cost of regulations imposed on the food chain.

Even for those who can afford it, the solution is not necessarily found in organic produce. A Consumers Union study found pesticide residues in 23 percent of organic produce versus 73 percent of conventionally grown produce (Brasher 2002, p. A3). Residue amounts were inconsequential in both types of produce, however.

Improved pesticides applied in ounces rather than pounds per acre, modern tillage techniques, best-management chemical application practices, and government regulations seem adequate to reduce currently low U.S. levels of synthetic chemical contamination of food and water to even lower levels. At the same time, the world much needs better tests for cancer-causing ingredients in foods.

The Centers for Disease Control and Prevention (CDC) estimate that 5,000 Americans die, 325,000 are hospitalized, and 76 million are made sick each year by food-borne pathogens such as salmonella. Thus, reducing pathogens in food is of higher priority than reducing pesticide residues for food safety.

**Water**

Over 70 percent of the earth's surface is covered by water, so plenty of water is potentially available for all people to meet their needs. Having water accessible at the local and regional level is primarily an economic and political issue. Americans use 1,442 liters of water per day per person. But humans need only about 50-100 liters per day for baths, laundry, and the like, including 2 liters of drinking water (Lomborg 2001, pp. 151, 152). Kuwait and Libya have less than 100 liters per capita per day of water from local sources, but fortunately they have petroleum to pay for desalinization of seawater.

Lomborg (2001, pp. 150-53) claims that desalinization could supply the earth with water for 0.5 percent of global GDP. He goes on to note that solar cells occupying less than 0.3 percent of the Sahara Desert could supply the power for desalinization. Such estimates, suggesting potentials, are not intended for practical application.

Practical water policies must consider agriculture. Globally, agriculture accounts for 69 percent, industry 23 percent, and households 8 percent of all water use. In most of the world, agriculture uses water very inefficiently. Many governments subsidize irrigation water to farmers, not charging enough to cover even the operating cost of public irrigation systems. Charging the full marginal costs of supplying water and emphasis on drip irrigation technologies can sharply raise water use efficiency.

With greater water use efficiency induced by improved technology and with full marginal cost pricing, opportunities abound to shift water of low use value in agriculture to consumers with high use value. The challenge is to establish markets or political institutions facilitating trade among farmers and consumers. Farmers and consumers alike can gain from that process.

If more food is needed in water-short-countries, that food can be imported from countries better endowed with water. Pursuit of sound policies for economic growth will ensure nations of buying power for water available domestically or elsewhere.

### Hypoxia

Hypoxia is not a major environmental problem but has attracted much interest in recent years. Hypoxia occurs as algae, seaweed, plankton, and other plants grow in coastal waters in response to nitrogen and (to a lesser extent) phosphate nutrients flowing from rivers. The nutrients enter rivers from farmland, urban lawns, and city sewage treatment plants. Eutrophication occurs as plant growth depletes oxygen in the water, killing marine animals that cannot leave for other areas or in other ways adapt.

Hypoxia takes place in numerous waters of the world but the best known is in the Gulf of Mexico at the mouth of the Mississippi River in an area typically the size of New Jersey. As human activity has expanded in the Mississippi-Missouri River basin, hypoxia appears to have increased.

The *Hypoxia Assessment* (Diaz and Solow 1999, pp. 8ff) estimates that a 20 percent fertilizer reduction coupled with 5 million additional acres of wetlands would entail a net additional annual cost of $2 billion. Fewer nutrients flowing into the gulf may decrease plant life and increase small animal life. But over the season and in the long run, less flow of nutrients is unlikely to increase animal life or commercial fish harvest. The benefit is mostly replacing small marine life with other small marine animal life.

The so-called hypoxia dead zone is actually teaming with life but not with the same organisms that would be there with less flow of nutrients into the gulf. Which set of organisms is preferred? Ultimately, the answer is a political judgment of whether benefits of reducing the flow of nutrients into the gulf justify net costs of some $2 billion per year for remedial policies.

Restrictions on fertilizer use are not costless. Farmers produce food and fiber more abundantly at lower cost and with less overall erosion (because they need fewer crop acres) when they use commercial fertilizer. And more acres of wetland "filters" can increase the mosquito population and mosquito-borne diseases troubling people and other animals.

Oxygen levels in rivers and other measures of water quality in the United States and other industrial nations are improving (CEQ 1997, p. 299). Nitrogen is frequently found in well water in the United States, but only very rarely at levels considered to be unsafe.

## COSTS OF RADICAL ENVIRONMENTALISM

A society serially misinformed about environmental risks can do foolish things. One reason is because limited public and private funds may be diverted from needed investments in health and human services to ineffective or counterproductive measures intended to "save" the environment.

Overreaction of environmentalists to perceived threats is illustrated by episodes with Alar, the spotted owl, the snail darter, and starving China. Readers may remember the spotted owl and snail darter cases; only Alar and the China cases are discussed here because they especially affected agriculture.

In the case of China, Lester Brown contended that the country is teaching that the western industrial model is not viable, simply because there are not enough resources (Brown 1995, p. 24). He concluded that China would starve poor countries. It would do so by annually absorbing over 200 million metric tons of grain imports by year 2030–a large share of the world's grain available for export. Brown effectively spread this message to East Asian countries that in turn redoubled their already considerable resolve for food self-sufficiency.

Reputable economists considered Brown's numbers to be absurdly high, but the damage had been done. The cost of his numbers is massive in lost national income to East Asian countries pursuing food self-sufficiency policies and to American and other food exporters in lost markets.

The chemical Alar was sprayed on trees to control apple ripening. Although federal regulators had cleared Alar for spraying on apple trees, consumers panicked when in the late 1980s the media raised alarms regarding the safety of consuming apples from sprayed trees. Apple consumption plummeted, losing apple growers tens of millions of dollars in sales. Even if Alar posed a significant health risk (and there wasn't evidence that it did), the phase out could have been handled without large economic losses.

Overly zealous regulations have other serious long-term costs to society. As noted earlier, excessive restrictions on pesticides to reduce cancer can increase cancer rates by raising production costs and consumer prices so that consumers cut consumption of fruits and vegetables high in antioxidants.

Proposed measures to reduce global warming are especially costly in lost national income. Instead of spending $100-200 billion per year to cut emissions under the Kyoto Protocol, the United States might more cheaply control emissions by increasing outlays for research to reduce the cost of solar and other clean energy. A carbon tax would restrain fossil fuel use and pay for the research.

Personal lifestyle choices — smoking, chronic overeating, and auto accidents– are the major sources of premature mortality in the United States. Measures to alleviate these and investment in health insurance and a longer school day and year for disadvantaged youth can have a higher payoff in well-being than public regulations to sharply curtail agricultural pesticides and carbon dioxide emissions. A Harvard study (Tengs et al. 1995, p. 371) calculated that, by reallocating public regulatory funds from their current use to the most cost-effective means of saving years of human life, some 600,000 life-years (about 60,000 full lives) could be saved each year. Additional regulation of agricultural pesticides was not one of the cost-effective means to save more lives.

**Eating Organic**

The nation spends over $8 billion each year for organic food. I applaud having a free enterprise system that allows consumers to express that choice in the market. The problem is that many consumers are drawn to organic food and to vegetarian or vegan diets by false claims by radical environmentalists. Consumers are told that conventional farm commodities are laced with poisonous farm pesticides and that such commodities lack nutrients and taste because they are produced on sterile soils made barren of essential nutrients by chemical fertilizers.

Organic foods cost more because they require more resources to produce. A 1990 survey (Batte , Forster, and Hitzhusen 1993) indicated that organic crop yields in Ohio averaged 25-30 percent below conventional farming crop yields. Accounting for lower cropping intensity, for green manure crops needed to build fertility, and for organic fertilizers brought in from elsewhere, I concluded that it takes nearly twice as many acres to produce a given crop output organically as conventionally. My results are consistent with those from a comprehensive national study at Texas A and M University by Knutson et al. (1989) of the impact on food production of ending commercial fertilizer and pesticide chemical use.

In a more recent review, Liebhardt (2001) concluded that on average over a wide range of conditions across the United States, organic yields averaged 95 percent those of conventional agriculture. The difference between organic and conventional yields found in the above studies depends heavily on what fertilizer and pesticide additives are allowed. Enough additives can bring "organic" yields up to conventional yields, but blurs the distinction between the two means of production. Organically producing today's crop output would take many more acres. Problems of soil erosion would be severe as cropping would be extended to marginal, often environmentally fragile soils. Additional mechanical cultivation required in organic production also can raise soil erosion and diminish opportunities for carbon sequestration possible with no-till farming.

None of the above numbers measure resource productivity. Market price is one measure of relative productivity for products with similar final-demand characteristics. In the Batte et al. study, organic price premiums ranged from 67 percent for corn to 135 percent for wheat. The premium carries to the retail level–consumers paid more in 1990 and continue to pay more for organic than for conventionally grown foods. Assuming no glitches in markets, that is a good indication that considerably more resources are required to produce each unit of organic foods, hence productivity is less than for conventional foods.

Diane Bourn of Otago University and colleagues in New Zealand reviewed approximately one hundred studies comparing organic foods with those grown using agricultural chemicals (see Kovach May 2002). Organic foods were found to be no healthier or tastier on average than other foods. (The researchers noted that the bulk of the studies–mainly from Europe but also from the United States and Australia-were poorly done, but the direction of any bias was not clear.) The authors claimed environmental benefits for organic foods, but did not consider the impact on the environment of soil erosion and wildlife losses from widespread low-yield organic cropping. People may prefer to eat organic for their own reasons, but they are unlikely to be doing the environment a favor.

**Violence**

Violence is costly. The radical environmentalists discussed thus far in the chapter probably abhor violence. But strong language and deeply held ideologies and animosities often move unbalanced individuals to violence. The Earth Liberation Front (ELF) claims "credit" for some 30 acts of violence to protect the environment, but may be responsible for many more. It says it is responsible for a 1998 fire that destroyed ski lifts and buildings in a Vail, Colorado, ski resort; a fire that destroyed an Indiana luxury housing complex; attacks on Washington and Oregon timber companies; firebombing a biogenetics office at Michigan State University; and setting fire to a number of sports utility vehicles (Kovach, February 2002, p. 1).

ELF announced that it was responsible for a blaze on January 26, 2002 that damaged heavy equipment and a trailer being used to construct the Microbial and Plant Genomics Center at the University of Minnesota's St. Paul campus. The laboratory will perform basic research on genetics to find plants that need fewer pesticides. The fire also damaged a nearby crop science laboratory. ELF said that it also was responsible for damage to a greenhouse at the university two years earlier (Kovach, February 2002, p. 1).

Issues deemed urgent by radical environmentalists are sometimes crosscutting with those of antiglobalists and Luddites. Chapters 1, 3, and 5 contain accounts of violence by antiglobalists and Luddites that could also be listed here because of ties of the individuals and organizations involved to radical environmentalism.

## CONCLUSIONS

Compared to previous generations, our generation lives longer. We also live better: we eat better, are healthier, have more leisure time, have more purchasing power, have more entertainment options, suffer less in the heat and cold, and face less drudgery on the job. Although poor nations lag, people there too are making progress and face ever lower incidence of poverty and hunger. The proportion of the world's people who rarely get enough to eat fell from 37 percent in 1970 to 18 percent in 2000 and promises to continue to fall (Andersen 2001, p. 4).

Progress is not coming at the expense of the environment. Technology and other forms of knowledge originating from human ingenuity and judicious investments in education and science have multiplied real output of goods and services available per unit of natural resources. Thus living standards can continue to rise indefinitely per capita even as natural resources are drawn upon.

Paradoxically, one source of this phenomenon is affluence. It provides resources to invest in research to raise productivity, to preserve and protect the environment, and to reduce population growth. All that makes the "system" much more than merely sustainable!

Agriculture too is more than sustainable. American agriculture supplies four times as much output today as a century earlier with only a little more aggregate real volume of resources. The success of agriculture is the result of teamwork between farm operators, agribusinesses, and the scientific establishment. The "team" would not have come together without a supportive institutional structure of regulations, investments in knowledge, and security provided by government. Soil erosion rates have dropped sharply since the 1930s. Chemical use on farms does not pose a threat to people through contamination of our water, air, or food. Limitations on supplies of land, water, energy, and minerals are unlikely to raise real food prices in the future.

All is not well, however. Regarding food, cod harvests in the north Atlantic are down; much grazing land is in poor condition. These circumstances do not threaten global food supply. They caution that open access property such as oceans and public lands will be overused unless a market or other institutional arrangement is created to control access.

Global warming and loss of biodiversity also are troubling to agriculture. An appropriate public response to global warming may be a tax on carbon emissions and research on clean energy sources such as nuclear and solar power. Creating and enforcing nature preserves and gene banks can reduce loss of biodiversity. Rising income in poor countries can provide the resources to increase agricultural productivity on "safe" soils and, eventually, reduce population growth so that poor people will not be pushed onto biodiverse but delicate ecosystems to avoid starvation.

The point of this chapter is that people can continue to live better, fuller lives for the long term while preserving the environment. Investment in and application of knowledge are critical for that outcome. Radical environmentalists who serially distort facts and use violence to achieve sustainability not only are unethical, they are counterproductive Cassandras.

## REFERENCES


Ames, Bruce, and Lois Gold. "Pesticides, Risk, and Apple Sauce." *Science* 244 (1989): 755-57.

Andersen, P. "Attaining Sustainable Food Production for All." Manshold Lecture, Wangeningen University, The Netherlands. Washington, DC: International Food Policy Research Institute, November 14, 2001.

Barton, P. "Renewable and Nonrenewable Resources." Presented to the American Association for the Advancement of Science meeting in Baltimore, Maryland, Feb. 10, 1996. Washington, DC: U.S. Department of the Interior, 1996.

Batte, M., D. L. Forster, and F. J. Hitzhusen. "Organic Agriculture in Ohio: An Economic Perspective." *Journal of Production Agriculture* 6, 4 (1993): 536-42.

Bos, E., M. Vu, E. Massiah, and R. Bulatao. *World Population Projections, 1994-95 Edition*. Baltimore, MD: Johns Hopkins University Press, 1994.

Brasher, P. "Study Finds Pesticides in Organic Produce." *Columbus Dispatch*. May 8, 2002, p. A3.

Brown, L. *Who Will Feed China: Wake-up Call for a Small Planet.* London: Earthscan Publications, 1995.

Carson, R.. *Silent Spring.* Boston: Houghton Mifflin, 1962.

CEQ. *Environmental Quality 1996.* The President's Council on Environmental Quality. http://www.whitehouse.gov /CEQ/reports/1996/toc.html, 1997.
Chakravorty, U., J. Roumasset, and K. Tse. "Endogenous Substitution among Energy Resources and Global Warming." *Journal of Political Economy* 105, 6 (1997): 1201-34.
Council of Economics Advisors. *Economic Report of the President.* Washington, DC: United States Government Printing Office, 1973 and later issues.
Crosson, P. "Cropland and Soil: Past Performance and Policy Challenges." In *America's Renewable Resources*, edited by Kenneth Frederick and Roger Sedjo. Washington, DC: Resources for the Future, 1992.
"Defending Science." *Economist,* February 2, 2002, pp. 15-16.
Diamond, J. *Guns, Germs, and Steel.* New York: W.W. Norton, 1997.
Diaz, R., and A. Solow. *Gulf of Mexico Hypoxia Assessment: Topic No.2. Ecological and Economic Consequences of Hypoxia.* Hypoxia Work Group, White House Office of Science and Technology Policy, Committee on Environment and Natural Resources for the EPA Mississippi River/Gulf of Mexico Watershed Nutrient Task Force. Washington, DC: NOAA Coastal Ocean Program. http://www.nos.noaa.gov/products/pubs hypox._html, 1999.
Ehrlich, P. *The Population Bomb.* New York: Ballantine Books, 1968.
Ehrlich, P., and A. Ehrlich. *Betrayal of Science and Reason: How Anti-Environmental Rhetoric Threatens Our Future.* Washington, DC: Island Press, 1996.
EPA (Environmental Protection Agency). *National Pesticide Survey: Summary Results.* Washington, DC: Office of Water and Office of Pesticides and Toxic Substances, EPA, 1990.
FAO (Food and Agricultural Organization). FAOSTAT Statistics Database, http://apps.fao.org, 2000.
Gore, A. *Earth in the Balance: Ecology and the Human Spirit.* Boston, MA: Houghton Mifflin, 1992.
Hervani, A., and L. Tweeten. "Kuznets Curves for Environmental Degradation and Resource Depletion." In *Agricultural Policy for the 21st Century,* edited by L. Tweeten and S. Thompson. Ames: Iowa State Press, 2002.
IPCC (Intergovernmental Panel on Climate Change). *Climate Change 1995–The Economic and Social Dimensions of Climate Change.* Report of IPCC Working Group III. Cambridge, UK: Cambridge University Press, 1996.
IPCC. *Climate Change 2001: The Scientific Basis.* Contribution of Working Group I to the Third Assessment Report of the IPCC. Edited by T. Houghton, Y. Ding, D. Griggs, M. Noguer, P. van der Linden, and D. Xiaosu. Cambridge: Cambridge University Press, 2001.
Knutson, R., C. R.. Taylor, J. Penson, and E. Smith. "Economic Impacts of Reduced Chemical Use." College Station, TX: Knutson and Associates.1989.
Kovach, J. "Ecoterrorists Set Fire to Minnesota School Lab." gmo@ag.ohio-state.edu, February 1, 2002.
Kovach, J. "GMO: NZ Questions Claims that Organic Foods are Tastier, Healthier." gmo@ag.ohio-state.edu, May 2, 2002.

Liebhardt, B. *Information Bulletin.* Davis, CA: UC-Davis, Organic Farming Research Foundation, Summer 2001.
Lomborg, B. *The Skeptical Environmentalist.* Cambridge, UK: Cambridge University Press, 2001.
Lutz, W. W. Sanderson, S. Scherbov, and A. Goujon. World Population Scenarios for the 21st Century," In *The Future Population of the World,* edited by Wolfgang Lutz. Laxenburg, Austria: International Institute for Applied Systems Analysis, 1996.
Magleby, R., W. Crosswhite, C. Sandretto, and C. T. Osborn. "Soil Erosion and Conservation in the United States: An Overview." *Agricultural Information Bulletin.* Washington, DC: Economic Research Service, USDA, 1995.
Malthus, T. *Essay on the Principle of Population.* Oxford, UK: Oxford University Press, 1798.
Myers, N. "Specious [sic]: On Bjorn Lomborg and Species Diversity." *Grist.* http://www.gristmagazine.com/grist/books/lomborg, December 12, 2001.
Mendelsohn, R., and J. E. Neumann, eds. *The Impact of Climate Change on the United States Economy.* Cambridge, UK: Cambridge University Press, 1999.
Paddock, W., and P. Paddock. *Famine–1975!* Boston: Little, Brown, and Company, 1967.
Parry, M., N. Arnell, M. Hulme, R. Nicholls, and M. Livermore. "Buenos Aires and Kyoto Targets Do Little to Reduce Climate Change Impacts." *Global Environmental Change* 8, 4 (1998): 285-389.
Pierce, F., and P. Nowak. "Soil and Soil Quality: Status and Trends." In *1992 Resources Inventory Environmental and Resource Assessment Symposium Proceedings, 1-23.* Washington, DC: NRCS, U.S. Department of Agriculture, 1994.
Rosenzweig, C., and M. Parry. "Potential Impact of Climate Change on World Food Supply." *Nature* 367 (1994): 133-38.
Schneider, S. H. "Hostile Climate: On Bjorn Lomborg and Climate Change." *Grist.* http://www.gristmagazine.com/grist/books/lomborg, December 12, 2001.
Schultz, K. "Let Us Not Praise Infamous Men: On Bjorn Lomborg's Hidden Agenda." *Grist.* http://www.gristmagazine.com/grist/books/lomborg, December 12, 2001.
Seldon, T. M., and D. Song. "Environmental Quality and Development." *Journal of Environmental Economics and Management* 27 (1994): 14-62.
Shafik, N. "Economic Development and Environmental Quality: an Econometric Analysis." *Oxford Economic Papers* 46 (1994): 757-73.
Simon, J. *The Ultimate Resource 2.* Princeton, NJ: Princeton University Press, 1996.
Smith, T. "Activists Blast Intellectual Property Rules." *The Desert Sun.* Palm Springs, CA, February 3, 2002, p. A17.
Solow, R. "On Intergenerational Allocation of Natural Resources." *Scandinavian Journal of Economics* 88 (1986): 141-49.

Stratton, L. "Afraid of the Water." *Columbus Dispatch*. March 24, 2002, pp. 11, 12.

Tanner, M. *Potential for Carbon Sequestration on Ohio Cropland.* M.S. Thesis. Columbus: Department of Agricultural, Environmental, and Development Economics, Ohio State University, 2001.

Tengs, T., M. Adams, J. Pliskin, D. Safran, J. Siegel, M. Weinstein, and J. Graham. "Five-Hundred Life-Saving Interventions and Their Cost-Effectiveness." *Risk Analysis* 15, 3 (1995): 369-90.

Tweeten, L. "Dodging a Malthusian Bullet." *Agibusiness* 14 (January/February 1998): 15-30.

Tweeten, L. "The Economics of an Environmentally Sound Agriculture (ESA)." In *Domestic and International Agribusiness Management*, Vol. 10, edited by Ray Goldberg. Greenwich, CT: JAI Press, 1992.

Tweeten, L., and W. Amponsah. "Sustainability: The Role of Market Versus the Government." In *Sustainability in Agriculture and Rural Development,* edited by G. E. D'Souza and T. G. Gebremedhin. Aldershot, UK: Ashgate Publishing, 1998.

U.S. Department of Agriculture. *Yearbook of Agriculture*. Washington, DC: U.S. Government Printing Office, 1938.

U.S. Department of Agriculture. *Agricultural Statistics*. Washington, DC: U.S. Government Printing Office, 2000.

UN (United Nations). *World Population Projections to 2150*. ST/ESA/SER.A/173. New York: UN, 1998.

UN. *World Population Prospects*. 2000 Revisions. http://esa.unpp/p2k0data.asp. New York: UN, 2002.

WCED. *Our Common Future* ("Brundtland Report"). The World Commission on Environment and Development for the General Assembly of the United Nations. Oxford, UK: Oxford University Press, 1987.

WI (Worldwatch Institute). *State of the World 2000*. New York: W.W. Norton, 2000.

Wigley, T. "The Kyoto Protocol: $CO_2$, $CH_4$, and Climate Implications." *Geophysical Research Letters* 25, 13 (1998): 285-88.

Wilson, E. O. "Vanishing Point: On Bjorn Lomborg and Extinction." *Grist.* http://www.gristmagazine.com/grist/books/lomborg, December 12, 2001.

World Bank. *World Development Report 1992: Development and the Environment.* Oxford: Oxford University Press, 1992.

# 5

# Luddites

## INTRODUCTION

America's most famous contemporary Luddite is Ted Kaczynski, the Unabomber. He was so distraught over the adverse impact of modern technology on society that he killed people to dramatize his cause.

The term "Luddite" originally referred to persons who destroyed laborsaving machinery being introduced into the British textile industry in the early nineteenth century. The terrorists (defined in this volume as those who unlawfully destroy people and/or property in peacetime) claimed they were acting on the orders of a mythical general Ned Ludd. In fact, the destruction was to save jobs in the face of mechanization that had accelerated with the industrial revolution which began about 1750.

Tactics and targets have changed over the years, but Luddites persist and are nowhere more apparent today than in agriculture. Decent, knowledgeable people often differ in their judgment of the merits and demerits of a particular technology. Such disagreement is appropriate and expected. Compared to conventional critics of technology, Luddites use unethical means such as deceit, intimidation, and violence to stop technology. Their vicious tactics have won transitory battles, but surprisingly few long-term victories in the wars against technology.

Modern-day Luddites are likely to oppose technology for reasons beyond retaining their jobs. The motivation may be to preserve the environment, protect the health of consumers, or save small farms. Modern-day Luddites also may be radical environmentalists, antiglobalists, and/or populists discussed in this volume. Hence this chapter sometimes overlaps with other chapters (Box 5.1).

**Box 5.1**
Sometimes the opposition to technology seems to come right out of the postmodern philosophy outlined in chapter 2. An academic philosopher, whom I was debating at an ethics symposium at Iowa State University held in 1986 regarding bioengineered bovine somatotropin, graciously conceded that I had adequately refuted his arguments against the hormone's use. But he added that he would find more reasons for prohibiting the hormone because in his words "It is ugly". That charge is difficult to refute. It seemed that he was not about to sacrifice his "gut feelings," whatever the opposition offered in facts, logic, science, and analytical rigor.

**Why Luddites Are Successful Today**

Global markets, large size, extensive interconnectedness, and a favorable image characterize today's successful multinational firm. The same communication and transportation technology that creates global reach and connectedness also creates economies of size for Luddite organizations. Activists skilled in fundraising and use of the media and information technology exploit vulnerabilities of firms depending on a favorable public image. Having to face only a few, image-conscious firms makes Luddites effective.

Schweikardt and Browne (2001, pp. 303, 313) relate how Greenpeace, an environmental and Luddite organization, conducted what Orden and Paarlberg (2001) call "politics by other means." The organization sent Gerber Products Company a fax in 1999 asking whether the company had taken steps to avoid using genetically modified (GM) ingredients in baby food. Gerber realized it must act quickly to avoid a public campaign by Greenpeace against the wholesomeness of Gerber products. Within days, major segments of the food industry, including Gerber, announced plans to limit ingredients from GM crops in their products. An irony is that Novartis, a major supplier of GM seeds to farmers, owned Gerber.

This example of what Miller (2001) called "turn tail and run" and Avery (2001a) called "corporate cowardice" in response to what Kava (1999) termed "food terrorism" by Greenpeace is not the only example of the success of Luddites. A tomato paste producer in Britain gathered a large market share on the strength of the quality of its genetically modified tomato ingredients. Greenpeace promoted use of the term "frankenfood" (a term apparently coined by the British tabloid press) for foods containing bioengineered ingredients. The pejorative fostered Frankenstein monsterlike images of the sinister, uncontrollable, grotesque, and dangerous. Luddites drove the tomato paste maker out of the market in the same way that other GM products, also safe for consumers and environmentally friendly, were driven out by a campaign of vilification.

Other Luddite organizations use other tactics. The Earth Liberation Front (ELF), labeled by the Federal Bureau of Investigation as "one of the leading terrorist organizations" (Fullerton 2001, p. 15), illustrates the problem of trying to classify a group. ELF has a footprint of predation ranging from destruction of housing sites in Long Island to lumber yards in Oregon. It has destroyed agricultural research laboratories and experimental plots involved in genetic engineering.

Given the controversial nature of recombinant DNA research, such destruction may not come as a surprise. But ELF also destroys conventional plant breeding facilities. An example is the May 2001 bombing of the Center for Urban Horticulture at the University of Washington. In response, President Bob Hoover of the University of Idaho stated "The ultimate goal of the terrorists–to use the threat of violence to limit scholarly inquiry–should be abhorrent to the entire academic community" (Fullerton 2001, p. 15). By limiting scientific inquiry, Luddites seek to stop technologies that benefit people and the environment even before those technologies are developed.

Modern biotech can become a target *or* a tool of terrorists. Jeremy Rifkin, who could be listed in any of the categories of radical or populist found in this volume, in an ironic twist proposed in the *Los Angeles Times* that GM research and technology be stopped because it could become a tool of terrorists spreading deadly organisms (Avery 2001b, p. 1). Most any worthwhile technology can be abused. The following pages make the case that GM technology offers far more promise than peril to society.

Luddites oppose a great many agricultural technologies. To save space, however, this chapter centers mostly on bioengineering and farm-scale technologies.

## BIOENGINEERING

Before examining opposition of Luddites to modern bioengineering in the food chain, I review a little of the genesis of genetic modification of farm crops and livestock. Bioengineering includes cloning, tissue culture, monoclonal antibodies, and other technologies, but this section is confined mainly to genetic modification through recombinant DNA.

### Genesis

Nature took billions of years to form the plant and animal kingdoms by moving genes *among* and *within* species in the process of natural selection and species evolution. Early farmers and herdspeople accelerated the process with purposeful selection of traits (genes) for domestication and enhanced performance of plants and animals. Conventional breeding was pretty much confined to moving genes *within* species.

The transgenic recombinant DNA technology of modern genetic engineering allows purposeful transfer of genes among species to quickly and efficiently select for desired traits. First-generation genetically modified (GM) products mostly benefited producers. Perceived lack of consumer benefits complicated the job of "selling" the new biotechnology to the public. Second-generation technology now mostly on the drawing board offers major nutritional and health benefits to consumers, hence likely will increase acceptance of modern biotech despite efforts of Luddites to sabotage the technology.

Chymosin, an enzyme that curdles cheese, was an early first generation success for GM organisms. Chymosin until the 1990s had been taken from rennin found in stomachs of slaughtered calves. Scientists at Pfizer Incorporated spliced cow genes for chymosin into bacterial cells. These cells grown in stainless steel fermentation vats provided plenty of chymosin to cheese makers at half the cost of conventional sources. Nearly everyone found much to like about biotech chymosin, and the product won approval from the Food and Drug Administration in March 1990.

Other biotech products were less acceptable to activists. Bovine somatotropin (bST), a growth hormone, is secreted naturally in cows. It has been bioengineered, as has chymosin, to be produced by GM bacteria. Monsanto put bioengineered bST on sale in February 1994. The hormone injected into lactating cows increased milk production by 10-20 percent. It is used primarily by large dairies able to practice the careful management required for a favorable economic payoff from bST. The biotech hormone proved to be less threatening to dairy cows and milk quality than biotech opponents predicted.

Calgene developed a FlavrSavr tomato that, compared to conventional tomatoes, delays softening, reduces spoilage, and maintains taste. Calgene submitted the gene marker in the tomato to the FDA for review in 1990 and in May 1994 received confirmation that the tomato was as safe and nutritious as conventional tomatoes. Because of negative publicity from GM opponents and because of marketing and production glitches, the GM tomato enjoyed only limited success.

The greatest first-generation biotech successes have been Roundup Ready soybeans and Bt cotton and corn. Roundup Ready varieties contain a gene inserted to make the plant resistant to the "burn-down" herbicide glyphosate. Bt varieties have a gene inserted from the soil microbe *Bacillus thuringiensis* that produces proteins protecting the plant from pests such as the cotton bollworm or corn borer.

Introduced in the mid-1990s, by 2001 GM varieties accounted for nearly two-thirds of the soybeans and cotton and for one-fourth of the corn planted in the United States. GM soybean growers claim savings of $20-50 per hectare and GM cotton growers claim savings of $65 per hectare. Savings are in fuel, machinery, labor, and pesticide costs. Because yields are only modestly affected, first-generation GM technology is unique–unlike most farm technologies it does not directly increase output to lower commodity prices and receipts. Hence, farmers retain economic benefits for an extended period. That outcome helped to push global GM crop area from 1.7 million hectares in 1996 to 43 million hectares in 2000. But lack of price and nutrition benefits has created fertile ground for Greenpeace and other Luddite groups to plant fears of GM foods in consumers.

Second-generation biotech will offer more to consumers and hence seems less likely to encounter intense opposition. In 2002, some 117 drug products and vaccines are being marketed, and more than 350 more are on trial to fight more than two hundred diseases (Kay 2002). Some of these pharmaceuticals, called "farmaceuticals," will be taken from farm crops bioengineered to grow the desired medications at low cost. An estimated 1,273 biotechnology companies employ 150,800 workers in the United States. Despite the best efforts of Luddites, the biotechnology "genie is out of the bottle."

The promise of modern bioengineering is staggering. Bioengineering potentially can "correct" genes that contribute to cancer, Alzheimer's, and

Parkinson's diseases. Genetically modified organisms offer promise of plants that will better tolerate the stresses of heat, frost, and drought, and of aluminum and salt in soils. GM crops will efficiently produce nutriceuticals, pharmaceuticals, plastic, paper, and a host of other products.

So-called golden rice is an example of the promise of second-generation GM organisms. Traditional rice lacks vitamin A. Up to two million children die each year and another fifty thousand go blind due to vitamin A deficiency (Elias 2001, p. 11). Those numbers can be reduced because scientists inserted two daffodil genes and one bacteria gene that turned conventional rice into golden rice producing vitamin A. In 2000, Monsanto and another 30 firms holding 70 bio-patents released golden rice to the public domain so that it will be available to all without royalty payments. Even with this assist from biotech firms, the (yet unproven) commercial success of golden rice will hinge on disease and pest resistance, yield, and consumer acceptance.

The rice genetic blueprint (genome) mapped by public research agencies and private companies was released to the public domain in 2002. By allowing researchers to target genes that are responsible for specific traits, the mapped genome will lower the time and cost of selecting and implanting genes to improve plant performance.

Monsanto also is developing a mustard seed that contains vitamin A in India, virus resistant potatoes in Mexico and Kenya, and virus resistant papaya in Southeast Asia and Hawaii. Such technologies, already credited with saving the papaya industry in Hawaii, will provide major second-generation GM crop benefits for poor countries.

**Real and Alleged Dangers of GM Technology**

Luddites charge that GM plants kill monarch butterflies, become or create super weeds, and pose threats to human health. Let us look at these issues, beginning with the StarLink fiasco that allegedly threatened human health due to a breakdown in regulatory procedures.

The StarLink variety of GM corn was approved by the Environmental Protection Agency. The Food and Drug Administration withheld approval of StarLink for human consumption while the variety was being tested for possible allergic reaction in humans. Aventis, the French parent company, elected to release the variety for use solely as animal feed while awaiting FDA approval for humans.

Although StarLink accounted for less than 1 percent of the 1999-2000 corn crop, it inadvertently mingled with other corn at grain elevators and other points in the grain delivery system, eventually contaminating nearly half the total harvest. The discovery of StarLink in Taco Bell taco shells prompted a recall in September 2000. Eventually, more than three hundred StarLink-tainted products were pulled from supermarket shelves. Aventis paid out millions of dollars to farmers and others to get StarLink out of the food system. Cleanup costs eventually totaled over $1 billion. The problem gained international significance when Japan and South Korea, the biggest foreign buyers of U.S. corn, rejected contaminated shipments of grain. Plantings of GM corn stalled if only temporarily.

Several lessons follow from this incident. First, confining a crop solely to feed use is very difficult in a system accustomed to using the crop for food and feed. To avoid a similar regulatory breakdown, the Environmental Protection Agency announced on March 7, 2001, that it would no longer approve a GM crop for feed use only. Second, the experience is somewhat reassuring, in that even a system failure did not cause allergy or other health threat to consumers. Finally, the StarLink fiasco humbled all parties and engendered a sense of caution regarding a regulatory system inevitably subject to human error.

Luddites and the press created an uproar against Bt corn after a Cornell University researcher found that Bt corn pollen sprinkled on milkweed leaves killed monarch caterpillars given nothing else to eat. An extensive two-year follow-up study by a consortium of federal, university, and industry scientists led by the Agricultural Research Service of the U.S. Department of Agriculture found that Bt corn posed no significant risk to monarch caterpillars or butterflies (Kaplan 2002). The monarch experience told something about media behavior: the threat to monarch butterflies was front page news for weeks; the later finding of no significant risk was hardly noted in the media.

A team of scientists at Imperial College in Britain, after monitoring four crops at 12 sites for ten years, concluded that GM crops are no more likely than conventional plants to become super weeds (Walker 2001, p. 1). Neither are GM crops more likely than other plants to cross-pollinate with wild relatives to become hybrid super weeds or invasive species. Cross-pollination may occur, however, where GM plants have close relatives nearby that are not self-pollinating.

Activists contend that genetic use restriction technology (GURT) or "terminator" GM gene technology "is universally considered the most morally offensive application of agricultural biotechnology, since over 1.4 billion people depend on farm-saved seeds" (RAFI 2000). Such condemnation is unjustified for several reasons. GURT does not force farmers to buy expensive seed each year–farmers can continue to use seed from traditional varieties. Many farmers will choose the GM seed, however, because it likely will yield enough to pay for buying new seed each year and still leave more crop available to market or consume at home.

So-called terminator, or suicide, seeds that are only viable for one planting are not new. High-performance corn and other hybrid crop seeds have been used with great success by millions of farmers for decades. Like hybrids, GURT seed sales each year compensate private firms for developing improved varieties, assuring a flow of improved seeds over time to raise farm productive performance and living standards. An ideal solution for a public good (firms cannot charge farmers enough to cover development costs) such as seed is for the public sector to develop improved varieties and deliver them at no cost to farmers. Public research is always woefully underfunded in poor countries, however. Hence raising agricultural productivity and living standards requires public *and* private investment to improve plant and animal genetics.

**Biotech Policies**

Are GM ingredients in our foods safe to eat? The answer is yes, according to the Food and Drug Administration (FDA), the American Medical Association, National Academy of Sciences, and the American Dietetics Association. The Environmental Protection Agency regulates to protect the environment against contamination from GM organisms. The regulatory process is somewhat simplified with the FDA decision made after years of study that bioengineered foods are not inherently different from other foods. Hence where there is "substantial equivalence" to conventional foods, GM foods do not require special regulations. Genes added to organisms are themselves proteins digested by humans like other proteins. Special permits are required only if a new GM product contains major nutrient or content changes such as higher levels of toxins or allergens. Procedures are being modified to make review by federal agencies mandatory rather than optional before GM products can be released by originating firms to the market.

Luddites and many other people opposed to GM technologies push for labeling of all GM products. Food firms would pass to consumers the high cost of identity preservation and mandatory labeling that specifies the type and quantity of GM ingredients in food. Because GM products are safe for people and the environment, mandatory labeling would do little more than teach people that they can ignore labels.

Organic foods do not allow GM ingredients so consumers already have the opportunity to buy non-GM foods. *Voluntary* labeling of non-GM foods and perhaps production processes (e.g., free range raised chickens) makes sense. That way, the cost of identity and quality preservation falls on the minority of the population willing to pay for a degree of food and environmental safety beyond that judged adequate by regulatory agencies.

Other countries have had more tumultuous introductions to the new biotechnology. Complex new technologies can never be fully explained to laypersons, and ignorance can be costly when consuming food. Thus, trust in regulatory agencies is central to avoid high costs from Luddite demagoguery. In Europe, lingering memories of regulatory failure for thalidomide and mad-cow disease created fertile grounds for Greenpeace and like groups to dominate the biotech dialogue. In 1986, Jeremy Rifkin, an American Luddite long opposed to gene manipulation, convinced Benedikt Haerlin, an official of the Green Party in Germany and later head of Greenpeace, to do everything possible to stop GM foods in Europe (McNeil 2000). The antibiotech campaign was so successful that American farm exports to Europe were severely curtailed at considerable cost to American farmers and taxpayers.

Europe might be rich enough to forgo GM foods, but one might expect the Luddites of Europe to be more supportive of the technology for poor countries. Mr. Haerlin of Greenpeace dismissed the importance of saving African or Asian lives at the expense of spreading "untested" GM technology.

An African official with the United Nation's Food and Agriculture Organization took a different line, observing that "organic farming is practiced by 800 million poor people in the world because they can't afford pesticides and fertilizers–and it's not working" (McNeil 2000). GM crops and livestock can improve farming productivity necessary to get low-income countries on the path to development. Agricultural productivity and living standard are two faces of the same coin for millions of poor, near-subsistence farmers.

Andrew Pollack (2001, p. 6), after observing that opponents of biotech were urging Kenya and India to reject GM corn and soybeans donations despite urgent humanitarian needs, asked "Are they [biotech opponents] so against it that they are willing to let people die?" The answer is "yes." Greenpeace is not all words; it acts, allegedly vandalizing GM corn and other crops across continents.

Opposition is not just from radical organizations. U.S. Congressman Dennis Kucinich from Ohio rejects the new green revolution of biotech seeds because the last green revolution of high-yield varieties from conventional breeding "didn't end hunger" (Mallaby 2001, p. 8). Most people would judge the first green revolution a huge success. *The Economist* ("Survey of Technology and Development" 2001, p. 4) says it saved a billion lives. In the region with the greatest number of undernourished people, South Asia, the first green revolution doubled cereal production and reduced the proportion of chronically undernourished people from 40 percent to 23 percent (Mallaby 2001). Such numbers remind us of the high cost of allowing Luddites to stop modern GM technology offering as much if not more promise than the first green revolution.

### International Regulation

Countries historically have used bogus claims that imported food is unsafe to justify protectionism. In a bold move, the Uruguay Round of multilateral trade negotiations completed in 1994 requires that sanitary and phytosanitary (health) or SPS restrictions on trade can be imposed only on scientific grounds.

The Europeans have sorely tested that provision. In meetings to consider revisions in the *Codex Alimentarius,* which codifies SPS regulations governing trade for members of the World Trade Organization, the Europeans pushed hard for labeling of GM foods while the United States fought successfully to avoid mandatory labeling. A U.S. concern was that Luddite and environmental groups could demonize safe foods with GM ingredients, causing consumers to shun safe foods while demanding and receiving protection from imports.

The Europeans were more successful in January 2000 when 130 nations including the United States meeting in Montreal hammered out an agreement to label international farm bulk commodity shipments. The agreement did not specify labeling of consumer food products in trade, however. Finally, Europeans and others can mandate GM labeling for reasons other than food safety, in which case challenges would not be filed under SPS but rather under the much weaker Technical Barriers to Trade Agreement (Sheldon 2002, p. 165).

## SMALL FARM TECHNOLOGY

Americans love the small family farm, although they have trouble defining what one is. Numbers of such farms by almost any definition have fallen sharply since the 1930s, but numbers have somewhat stabilized in the past decade. The combine, the tractor and its complements, the milking machine, and numerous other laborsaving machines have caused farms to get larger. Given the rather fixed land in farms, that means that farms and farm population have gotten fewer.

At the core of farm size, numbers, and consolidation issues is the concept of *economies of farm size*–the total cost per unit of production on farms of various sizes. That cost per unit includes operating costs as well as the annualized value of overhead capital inputs. Lower unit costs for larger operations cause farms to grow in size and decline in numbers.

Capital items that save labor and other inputs are often expensive, "lumpy," and durable. To reduce the average overhead cost, these machines must be used to produce many units of output each year. Economies of size drive farm structure, providing incentives for operators to cut costs per unit by consolidating farms into larger operations. This process frees persons on farms to work in industries favored by society as wealth grows. Mechanization also allows small farms to survive by cutting farm work time so that operators and spouses can supplement their income with off-farm employment. Stated differently, mechanization allows nonfarm workers to operate farms in off-hours while enjoying the attractive amenities of rural residency.

A massive literature documents compelling evidence of economies of size (Tweeten 1984; Hallam 1993). Compared with the actual food costs, Tweeten (1984, p. 49) calculated that reliance solely on small farms would raise food costs 14 percent, reliance solely on medium-sized farms would raise them 7 percent, and reliance solely on large farms would reduce food costs by 4 percent in the early 1980s. Differences in unit costs between large and small farms have not narrowed since that study. Consumers would pay a high price for a policy allowing only small farms.

With less than 1 percent of the nation's workforce required to supply food and fiber, the other 99 percent have had time and resources to find cures for diseases, and to educate, entertain, clothe, and house the nation. Meanwhile, real incomes of farm and nonfarm people alike have reached unprecedented levels.

So who could fault the technology making such progress possible? Wendell Berry, a favorite of agricultural Luddites, in 1977 observed

> the connection between the "modernization" of agricultural techniques and the disintegration of the culture and the communities of farming–and the consequent disintegration of the structures of urban life. ...The aim of bigness implies not one aim that is not socially and culturally destructive. And this community-killing agriculture, with its monomania of bigness, is not primarily the work of farmers. ...It is the work of the institutions of agriculture: the university experts, the bureaucrats, and the "agribusinessmen" who have promoted efficiency at the expense of the community. (pp. 41, 42)

Critics could accuse Berry, an English professor, of wordiness but not of inconsistency when in 2002 he reiterated that

> the industrial and corporate powers, abetted and excused by their many dependents in government and universities, are perpetrating a sort of economic genocide–less bloody than military genocide, to be sure, but just as arrogant, foolish, and ruthless, and perhaps more effective in ridding the world of a kind of human life. The small farmer, and the people of small towns are understood as occupying the bottom step of the economic stairway and deservedly falling from it because they are rural, which is to say not metropolitan or cosmopolitan, which is to say socially, intellectually, and culturally inferior to "us". (p. 4)

Berry (1977, pp. 44, 45) lamented that "from a cultural point of view, the movement from the farm to the city involves a radical simplification of mind and of character. ...Nor is there any acknowledgement of the influence of 'monster' technology ('acre-eaters') on the soil, the produce, the farm communities, and the lives and characters of farmers."

There you have it. Modern farming technology has ruined almost everything including the minds of city people according to Berry. The solution claims Berry (pp. 203-20) is a return to horse draft power, self-sufficiency, mandatory supply management, and, in general, a heavy hand of government to dictate farm size.

Perhaps Berry's considerable ability to attract disciples is prima facie evidence of how seriously agricultural technology has damaged minds (Comstock 1987; Blatz 1991). At any rate, his following seems reason enough to devote a few paragraphs of response to Berry.

**Demystifying Small Farms**

In addition to having high production costs per unit, small farms have other shortcomings. One shortcoming is that they don't seem to provide their operators and families with the promised superior life–their occupants aren't happier than people elsewhere. Sociopsychological measures of well-being are of similar magnitude for persons in farm and other sectors of society (Drury and Tweeten 1998), and between operators of large versus small farms (Coughenour and Tweeten 1986).

One of Berry's myths, often perpetuated by the press, is that the environment is damaged more by big farms than by small farms. In a *New York Times* article headlined "Big Farms Making a Mess of U.S. Waters, Cities Say," Becker (2002, p. 16) attributes to Professor Michael Duffy of Iowa State University the statement that "the relationship between federal subsidies and water problems begins with farm payments that encourage big farms to grow bigger, buying out small farmers who tend to be better conservationists." One problem with this statement is that commodity programs provide considerably higher subsidies per dollar of output to small and medium sized farmers than to large farmers, hence provide little or no incentive for large farms to cannibalize their neighbors (Tweeten 1993). In 1999, large farms with sales of $250,000 or more accounted for over two-thirds of farm crop and livestock sales but received less than half of government payments (U.S. Department of Agriculture 2001, appendix 1).

A second problem with Duffy's statement is that, compared to small farmers, large farms on average are as good or better stewards of the environment. Large farms have a higher percentage of their crops in conservation tillage or no-till practices that save soil and on average give lower soil erosion rates than found on small farms (Tweeten 1995). Large farms are more likely to store manure so that it does not have to be spread on frozen soils where it is prone to run into waterways after a thaw or rain. Large farms are more likely to inject manure into the soil for fewer runoff and volatilization problems, and are more likely to follow an approved waste disposal plan (Tweeten , Harmon, and Feng 1999). And as indicated in chapter 6 on animal rights, large farms are more likely to achieve size economies with waste disposal equipment that could be mandated as more stringent environmental regulations are imposed on all farms.

Berry faults land grant colleges for loss of family farms. The loss, however, traces much more to mechanization technology originating from private firms than from public institutions. An estimated 85 percent of land grant agricultural college research is on biological and other scale-neutral technologies, and more of the remaining research is targeted to help small farms than large farms (Tweeten 1983, p. 39). More timely data are not available, but land grant colleges continue to emphasize scale-neutral research.

Initial and final scale neutrality are not the same, however. Technologies such as bST or computers that would appear to be scale neutral ultimately are scale biased. Bovine somatotropin works equally as well on a cow on a big farm or a small farm. However, profitable and productive "scale neutral" new technologies tend to be adopted quicker and more widely on larger farms. Better managers are attracted by monetary rewards on larger farms, and such managers seek out better ways of doing things. Thus, large farms are more likely than small farms to use forward pricing, computers, e-commerce, precision farming, and many other technologies that allow them to produce and market at lower cost and higher returns.

Even large farms are small by nonfarm business standards, and don't have much market power. The economic advantage of larger farms comes more from production economies and real marketing economies (ability to purchase inputs and sell outputs in volume) than from pecuniary marketing economies (market power).

Modern farm technologies and institutions often help small farms. Many small operators produce organically for local farmers' fruit and vegetable markets. Programs of land grant colleges have helped such operators. Large numbers of small, low-income farms were able to survive and prosper in the South because of broiler production contracts and the technology that attended the contracts. Production contracts allow small farms to survive financially by utilizing as many production modules as meet their needs.

For example, a small farm with a little extra labor and in need of income can utilize one or more barns, each finishing one thousand pigs. Each barn requires about one hour of labor per day. At the same time, that grower's vertically coordinated production contractor utilizes larger scale farrowing and feed or pork processing facilities to achieve economies of size essential to survive in the highly competitive hog business.

Despite compelling evidence of economies of size in livestock production even when externalities are internalized, some communities and states seek to guide technology and rules so as to maintain only small, traditional livestock operations. Because such operations require more inputs per unit of output (are less efficient) and because the operators tend to buy locally, local communities see merit in that small-farm system. The problem is that these small operations are unable to survive in the market without off-farm income to support their farm.

Luddites especially support technology for small farms. "Appropriate" technology, a buzzword of the 1970s, presumably would make small farms cost-competitive in markets. The concept has flaws. First, it is difficult for land grant colleges of agriculture or agribusinesses to come up with technologies serving small farms that do not serve larger farms even better. Almost all profitable technologies ultimately if not initially save at least some labor and tend to result in fewer small farms.

Second, public and private investments in technology have high economic payoffs because they reduce food costs. Small farms defined as those with sales of less than $100,000 per farm accounted for 84 percent of farms but for only 15 percent of farm crop and livestock sales in 1999 (U.S. Department of Agriculture 2001, appendix 1). Technologies confined to such a small market tend to have low economic payoff. Direct government payments targeting solely small farms would be more effective in saving and creating small farms, but Congress has shown little interest in payment limits without loopholes to neuter the policy.

Luddites lament the impact of modern technology on farms partly because loss of small farmers diminishes rural communities. The situation is not as dour as often portrayed. From 1990 to 2000, rural (nonmetropolitan) counties grew fully 14 percent while the U.S. population in total grew 10 percent. The rapid growth in rural communities could occur because most do not depend on a farming base. In 1999, only 258, or one-eighth, of all U.S. counties were classified as farming-dependent, deriving over 20 percent of their income from farming. In 1969, 877 counties were farming-dependent (U.S. Department of Agriculture 2001, p. 89).

Small towns in farming-dependent counties have declined for reasons besides farm mechanization. Farmers and small-town residents alike have utilized modern transportation and improved roads to travel to and shop in larger centers offering a greater variety of goods and services at lower cost to consumers.

The notion that governments can direct technology and laws to shape an ideal farm size is a conceit. For example, in East Asia governments undersized farms; in the former Soviet Union government oversized farms. A case can be made that midsized rather than small farms are best for communities. Not everyone can be a middle-class, midsized farm operator, however. It appears that markets are in a better position than governments to decide what size of farm is "best."

**Serving Equity**

Luddites favor restraining modern technology to serve equity, among other goals. Other radicals favor land reform to serve equity. Governments have more promising and productive means than technology policy or land reform to serve equity, however. In 1990, 85 percent of U.S. national wealth was human capital and only 4 percent was farm real estate. Even in low-income Africa and Asia, in 1994 human capital comprised 60 percent and "natural" capital including farmland comprised only 10 to 20 percent of all wealth (Dixon and Hamilton 1996). Thus, public education and other programs to promote broad-based economic development offer much greater opportunity to serve equity than do technology and landownership restraints.

## CONCLUSIONS AND POLICY OPTIONS TO ACCOMMODATE TECHNOLOGICAL CHANGE

Because economic payoffs from investments to improve agricultural productivity have been very high to society, forgoing such investments in deference to Luddites seems unthinkable. Indeed, the rise of civilization and of quantity and quality of human existence has been largely a product of human and material capital formation–enabling ever-lower proportions of the population to supply food needs. Not all technologies benefit society, and some have proposed that new technologies receive more public scrutiny prior to their release.

**A Prerelease Socioeconomic Impact Statement?**

Technologies must pass the private market test for success, but the following examples argue against requiring a socioeconomic impact statement prior to release of technology much as an environmental impact statement is required before a dam can be constructed. One example is from a seminar in 1966 at the Delhi School of Economics in India to anticipate the consequences from introducing green revolution seeds. David Hopper (1978, p. 69) reported on the outcome of a seminar of government bureaucrats, scholars, technical experts, foreign advisors, and India's minister of agriculture: "Despite protests of the few, the meeting carried a clear consensus for prohibiting the entry and use of the new [high-yield, green-revolution] varieties. Fortunately for the nation's hungry masses, the politicians ignored the consensus."

A second example is from a lawsuit filed by California Rural Legal Assistance attorneys (CRLA) against the University of California on behalf of workers who might be displaced by the tomato harvester machine and other laborsaving devices being developed by the University (Martin and Ohlmstead 1984, p. 25). CRLA argued that mechanization research displaced farm labor, eliminated small family farms, diminished the quality of rural life, and harmed consumers.

The third example was an effort by the Foundation on Economic Trends and the Humane Society to stop the National Institutes of Health (NIH) from financing research that transferred genes from one mammalian species to another (McDonald 1984, pp. 7ff). The two groups asked that NIH cut off all support for any institution engaged in such research. The same two groups filed a lawsuit against the U.S. Department of Agriculture to halt a study to transfer human growth hormone genes into pigs and sheep.

These Luddite efforts failed. The technologies they tried to stop have vast actual or potential benefits. Critical ethical issues must be addressed, and the above examples indicate that even learned people are not very good at sorting out what is useful. The conclusion is that careful calculation of benefits and costs of technology can help, but alone is inadequate for sound decisions. Prudent application of regulations, taxes, subsidies, and other means to address externalities of technologies is important, but also is inadequate. Extensive dialogue among technology stakeholders over ethical, social, economic, and political implications of major new technologies is helpful, but that too is inadequate for sound decisions. To create trust, all these measures need to come into play.

**Improving the Social Contract**

A successful society runs on trust, and that elusive concept must be cultivated with care. As with international trade, technological change leaves some people behind. The unemployment rate is not much different between societies with fast and slow rates of technological change. While technology does not have much impact on the number of people working in society, it does at least temporarily unemploy some workers. Technology improves overall quality of jobs, but those displaced by technology can experience hardship as they adjust to new jobs.

A worthwhile technology produces enough dividend of net benefits so that those displaced do not need to be left behind for extended periods. A nation as affluent as the United States has plenty of resources to help persons adjust to change whether they are displaced by trade or technology.

How does the social contract apply to agricultural technology (see also chapter 10)? Commodity programs attempt to compensate farmers for technological change by suppressing farm-resource adjustments at costs in lost national income of $6.5 billion and in taxpayer outlays of $20 billion per year (Tweeten 2002, p. 12). Congress seems bent on continuing such efforts, however ineffective, to preserve farms. The total present cost to taxpayers in perpetuity is $400 billion when annual costs are discounted at 5 percent.

How many farms do commodity programs save? The U.S. Department of Agriculture (2000, p. 20) estimates that 4 percent of farms are financially vulnerable to failure, where "vulnerable" is defined as farms with negative cash flow from farming and a debt-asset ratio exceeding 40 percent. This definition overestimates the proportion of farms that would fail without programs because it omits all-important off-farm income and includes farms that will fail even with programs.

Let us be expansive, however, and assume that as many as 10 percent of farms would fail from technological change in the absence of programs. Thus, the cost to taxpayers averages $200,000 for each farm saved in perpetuity. It follows that the public could afford to pay up to $200,000 per farm for education and other adjustment assistance rather than continue current programs. In fact, much less spending per farm could provide generous adjustment assistance while strengthening the social contract by making consumers and taxpayers as well as farmers better off with technological change.

**Technology for Poor Countries**

Thousands of activists gathered in Porto Alegre, Brazil at a World Social Forum to counter the World Economic Forum meeting at the same time in February 2002 in New York City. Demonstrators charged that the international system of patents, copyrights, and trademark protections favor the rich countries where most of them originate over poor countries that cannot afford access to such intellectual property. Further, the argument ran, those varieties in many cases were mere bioengineered tweaking of the genes in native plants taken from poor countries.

In 1995 the World Trade Organization established rules for an international treaty system granting temporary monopolies to patent holders, many of them private firms, for varieties of plants, defined by their genes, they develop. Several nations had already established such rights; the World Trade Organization extended such rights to all members.

Poor countries such as Brazil and Argentina have benefited greatly from plant varieties developed by private firms, many of them multinationals headquartered in industrial countries. Few local farmers have complained. Farmers in poor countries have access to traditional varieties that are still available. Private firms have tended to provide patented varieties to farmers at lower cost in poor than in rich countries because demand is much more sensitive (elastic) to price in poor countries.

The economic base of most poor countries is agriculture. Those countries desperately need modern, high-yield technology, Luddite arguments notwithstanding. Conceding to activists' demands to curtail intellectual property rights and reject GM technology would relegate poor countries to extended poverty. In addition to fostering movement of commodities and intellectual and material capital among countries through freer trade, industrial countries can assist poor countries in a major way by public investment in basic research that ultimately will underlie the next green revolution. Sharing the benefits of basic and applied public agricultural research with all poor countries makes much more sense than trying to compensate a few poor countries for the genes that were borrowed from their public domain.

**REFERENCES**

Avery, D. "When Gerber Goes Organic, Fear Mongering Wins." Soybean.Com's Biotech Education Series. http://www.soybeans.com/avery 140.htm, 2001a.

Avery, D. "The Tomato that Ate Chicago." Press release. Washington, DC: Hudson Institute, October 5, 2001b.

Becker, E. "Big Farms Making a Mess of U.S. Waters, Cities Say." *New York Times,* February 10, 2002, p. 16.

Berry, W. *The Unsettling of America: Culture and Agriculture*. San Francisco: Sierra Club Books, 1977.

Berry, W. "The Prejudice Against Country People." *The Progressive*. http://www.progressive.org/April%202002/berry0402.html, 2002.

Blatz, C., ed. *Ethics and Agriculture*. Moscow: University of Idaho Press, 1991.

Comstock, G., ed. *Is There a Moral Obligation to Save the Family Farm?* Ames: Iowa State University Press, 1987.

Coughenour, C. M., and L. Tweeten. "Quality of Life Perceptions and Farm Structure." In *Agricultural Change*, edited by Joseph Molnar. Boulder, Colorado: Westview Press, 1986.

Dixon, J., and K. Hamilton. "Expanding the Measure of Wealth." *Finance and Development* 33 (December 1996): 4-7.

Drury, R., and L. Tweeten. "Have Farmers Lost Their Uniqueness? Response." *Review of Agricultural Economics* 20, 1 (Spring/Summer 1998): 206-7.

Elias, P. "Debate Grows on Golden Rice." *Chicago Tribune*, June 26, 2001, p. 11.

Fullerton, J. "Terrorists Among Us." *Farm Journal*, December 2001, pp. 14, 15.

Hallam, A., ed. *Size, Structure, and the Changing Face of American Agriculture*. Boulder, CO: Westview Press, 1993.

Hopper, D. "Distortions in Agricultural Development Resulting from Government Prohibitions." In *Distortions in Agricultural Incentives*, edited by T. Schultz. Bloomington: Indiana University Press, 1978.

Kaplan, K. "Bt Corn Poses No Significant Risk to Monarchs." News Release. Washington, DC: Agricultural Research Service, U. S. Department of Agriculture, February 6, 2002.

Kava, R. "Editorial: Fear Terrorists Targeting U.S. Parents." American Council on Science and Health. http://www.acsh.org/press/editorials/greenpeace082399.html, 1999.

Kay, A. "Biotechnology on Verge of Job Explosion." *The Desert Sun,* February 6, 2002, p. E3.

Mallaby, S. "No High-Tech Food for the Low-Income Hungry?" *International Herald Tribune*, July 12, 2001, p. 8.

Martin, P., and A. Olmstead. "Sprouting Farm Machinery Myths." *Wall Street Journal*, May 14, 1984, p. 25.

McDonald, K. "Attempts to Halt Genetic Research Anger Scientists." *Chronicle of Higher Education*, October 24, 1984, pp. 7ff.

McNeil, D., Jr. "New Genes and Seeds: Protesters in Europe Grow More Passionate." *New York Times*, March 14, 2000.

Miller, H. "Global Food Fight." *Hoover Digest*. http://www-hoover.stanford.edu/publications/digest/001/hmiller.html. Stanford, CA: Hoover Institution, Stanford University, 2001.

Orden, D., and R. Paarlberg. "The New Century of Multi-Agriculturalism." *Review of Agricultural Economics* 23, 2 ( Fall/Winter 2001): 289-301.

Pollack, A. "A Food Fight for High Stakes." *New York Times*, February 4, 2001, p. 6.

RAFI. "Suicide Seeds on the Fast Track." Winnepeg, Manitoba: International Office. http://www.rafi.org, March 2000.

Schweikardt, D., and W. Browne. "Politics by Other Means: The Emergence of a New Politics of Food in the United States." *Review of Agricultural Economics* 23, 2(Fall/Winter 2001): 302-18.
Sheldon, I. "Regulation of Biotechnology: Will We Ever 'Freely' Trade GMOs?" *European Review of Agricultural Economics* 29 (2002): 155-76.
"Survey of Technology and Development." *The Economist*,November 10, 2001, pp. 3-16.
Tweeten, L. "The Economics of Small Farms." *Science* 219 (4 March 1983): 1037-41.
Tweeten, L. "Causes and Consequences of Structural Change in the Farming Industry." NPA Report No. 207. Washington, DC: National Planning Association, 1984.
Tweeten, L. "Government Commodity Program Impacts on Farm Numbers." In Arne Hallam, *Size, Structure, and the Changing Face of American Agriculture*. Boulder, CO: Westview Press, 1993.
Tweeten, L. "The Structure of Agriculture: Implications for Soil and Water Conservation." *Journal of Soil and Water Conservation* 59 (July-August 1995): 345-49.
Tweeten, L. "Farm Commodity Programs: Essential Safety Net or Corporate Welfare?" In *Agricultural Policy for the 21st Century*, edited by L. Tweeten and S. Thompson. Ames: Iowa State Press, 2002.
Tweeten, L., C. Harmon, and X. Feng. "Independent and Contract Swine Producers' Attitudes towards Industrialization and Economic Change." Anderson Chair Publication ESO 2553. Columbus: Department of Agricultural, Environmental, and Development Economics, The Ohio State University, 1999.
U.S. Department of Agriculture. *Agricultural Income and Finance*. AIS-75. Washington, DC: USDA, September 2000.
U.S. Department of Agriculture. *Food and Agricultural Policy*. Washington, DC: USDA, September 2001.
Walker, A. "GM Crops Pose Few Ecological Risks." CNN interview. gmo@ag.ohio-state.edu, February 7, 2001.

# 6

# Animal Rightists

## INTRODUCTION

The human race has survived and prospered by using animals for draft power, clothing, fuel, transport, weapons (warhorses), fertilizer (manure), recreation (pets), and food. Especially in more recent times, animals have helped the handicapped, developed surgeon's skills, tested new medications including inoculations, and searched for illegal drugs. Because of the human debt owed to animals, it is altogether proper to raise social consciousness of them and treat them well.

Because animals lack capacity for higher cognitive language and thinking that characterizes humans, they are not in a position to formally protest cruel treatment. People speak for animals. According to an Associated Press poll, most Americans believe that animals have the right to a live free of cruelty (D'Amico 2001, p. 1). However, perhaps because animals are incapable of anticipating their slaughter and hence do not suffer fear of slaughter, most Americans don't consider animal slaughter and consumption to be cruel. American culture and diets are omnivorous. Those who would make the nation herbivorous are said to be counterculture activists. Vegetarians, vegans, and animal rights activists comprise a small percentage of the population.

Neither I nor most other Americans favor bestowing full human rights on animals. Granting animals full human rights would preclude all uses listed in the opening sentences. Even keeping animals for pets is slavery! However, people are entitled to their beliefs. Hence, beliefs are not the principal issue in this chapter. Rather, the issue here as in other chapters is use of unethical means such as misinformation and violence by animal rightists to achieve their ends–animal rights. Too often, animal rightists, a minority, use violence to try to impose their will on the majority.

Traditional societies cared much more about having enough to eat than the processes giving rise to the food. In contrast, affluent western societies care a good deal about processes of producing food. One reason for the change in attitude is because status in modern society is more about how one consumes than about what one does for a living. In traditional societies, family and neighbors observed one's daily labors in the fields and barns. People were held in high esteem for exhibiting excellence in their daily work.

How different today. Americans commute to their work in large office complexes far from relatives and neighbors. Their esoteric jobs are difficult to explain or too remote to earn status from friends, neighbors, and family. With job prestige defying definition, a person's worth, dignity, and status are judged by what can be observed–consumption lifestyles. Americans are judged by what they consume, including the processes by which food is produced.

As incomes rise, people consume more meat. But at some high level of income, some people cut back on meat consumption. Vegetarianism in the United States is rare among lower socioeconomic groups. The affluent are especially concerned about animal care and possible presence of drugs, antibiotics, and pathogens in meat. Consequently, they are willing to pay more for meat that is certified drug free and organic. These changing concerns and preferences have not offset forces for greater consumption: red meat and poultry consumption in the United States increased 21 percent in the 1990s and totaled 76 billion pounds in 2000 ("Absolutely Nuts About Numbers?" 2002, p. A6).

Certification, labeling, and offering people choices in supermarkets can allow people to go beyond customary federal regulations and agency inspections to ensure food safety. Even that wide array of options for food safety and choice is not enough for some people. They want to ban use of animals to benefit humans. This chapter addresses animal rights issues raised by these individuals, many of whom are vegetarian or vegan, the latter rejecting dairy and poultry products as well as animal flesh.

## GENESIS

Attitudes regarding use or abuse of animals in Western cultures have evolved from near universal adherence to the Judeo-Christian tradition to much more complex and varied attitudes in more recent years. This section examines these changing views.

### Judeo-Christian Tradition

The Judeo-Christian tradition dominating Western culture emphasizes the primacy of people over nature. Although animals are for the service of people, they are not to be abused. The first chapter of Genesis (1:24-28) states, "So God created man in his own image...[and said to man] be fruitful, and multiply, and replenish the earth, and subdue it, and have dominion...over every living thing that moves upon the earth." Other passages in the Old Testament call for kind treatment of animals and makes clear that the word "dominion" refers to "stewardship." Thus, the Judeo-Christian tradition does not condone abuse, misuse, or waste of natural or biological resources. Partly because only humans were regarded as having a soul, however, people afforded themselves a higher status than other animals. Furthermore, nature itself provided a hierarchy: micronutrients in the soil fed grass that fed grazing animals that fed carnivores that fed, clothed, and entertained humans.

### From Animal Welfare to Animal Rights

Not everyone subscribes to the Judeo-Christian position regarding animals. Those concerned with how animals are cared for and used can be classified into three broad categories: (1) animal welfarists, (2) animal welfare activists, and (3) animal rightists. Most farmers and most members of the agricultural establishment desire to see animals treated well, hence are *animal welfarists.*

Animal welfarists tend to rely on results from scientific research and information from traditional communication media to motivate and initiate changes in animal-care rules and regulations through the political process. *Animal welfare activists* feel acute urgency for animal-care reform, and frequently use demonstrations and other nonconventional, nonviolent means as well as conventional means to bring animal-care reform.

Finally, *animal rightists* would extend human rights to animals. Animal rightists reject the utilitarian benefit-cost analysis of the first two groups, and feel that urgency to protect animal rights justifies peddling of misinformation, hate, and, sometimes, violence to accomplish reform. The importance of the end justifies almost any means, lawful or unlawful, to bring justice to animals.

## ETHICS OF ANIMAL WELFARE

Economists like to employ a social welfare function to conceptualize what they do. That function expresses the well-being of society as a function of all goods and services and everything else that influences satisfactions of people. The welfare-maximizing allocation of goods and services can, in principle, be calculated from that function–if quantified.

At issue is who and what are included in the social welfare function! Is it only people whose well-being is being maximized? Do animals figure in only to the extent they contribute to the well-being of people? Or do animals (and indeed all of nature) have intrinsic value above and beyond the utility derived from them by humans? If animals have intrinsic value, then the social welfare function needs to maximize well-being of people *and animals* subject to resource and technology constraints.

Making final judgments about such issues is an exercise in values and politics rather than in science. Numerous individuals have grappled with this issue. The animal rights movement was energized by Peter Singer with his book *Animal Liberation* in 1975.[1] The movement seemed to be a logical extension of the contemporaneous black, gay, and women's liberation movements. The core principle of Singer's book (1990 edition, p. 243) is that "to discriminate against beings solely on account of their species is a form of prejudice. It is immoral and indefensible in the same way that discrimination on the basis of race is immoral and indefensible."

Drawing on moral philosophy rather than economics, Peter Singer (and the animal rights movement) found the answer to the question of animal rights to be clear: "All animals [man included] are equal" (1990, p. 1). He recognized that the black, gay, and women's liberation movements were all about people–beings who could understand, debate, communicate, and cast ballots about issues. Singer argued that equality for animals does not require equality of treatment. Rather, he argued that equality requires equal *consideration* (p. 2). That consideration could lead to different treatment of animals than of people.

Singer contended that the principle of equality requires *consideration* of all interests of "equal" members of society but does not depend on attributes such as mental ability of a being (p. 5). Thus, a pig is treated differently than a child not because their IQs differ but because a pig needs adequate room to run freely and the child needs schooling. Thus, all sentient species have rights but how those rights are expressed depends on the respective "interests" of the various species and subspecies. Conventional agriculturalists view such sophistry as mere verbiage, as one person's opinion.

So who should be granted inclusion in the social welfare function? Singer contends that all sentient beings should be included, that is, animals with a capacity for suffering and enjoyment (p. 7). Thus, Singer's position bears superficial similarity to the utilitarianism of Jeremy Bentham (see chapter 2) whose moral philosophy was that public policies be judged good or bad based on their contribution to pleasure and pain; the greatest good for the greatest numbers. The difference between Bentham and Singer is that the latter brings animals into the utility "equation."

Singer contends that "most human beings are speciesists" (p. 9). That is, they do not afford all beings that are similar in other respects the same rights. It comes as no surprise that his concern was lack of rights for animals (nonhuman). He rejected emphatically the proposition that economic reward to farmers and a good life for farm poultry and animals go hand in hand (p. 106). He lamented that the Animal Welfare Act does not apply to farm animals but was pleased that it applied to zoos, circuses, pet dealers, and laboratories (p. 111). Of course, farmers are subject to laws prohibiting cruelty to animals. And few people would deny that the loss in income from lower productivity of abused animals is a major goad to proper farm animal care.

Circumscribed by the positive right of animals to their "nature" (to have sex, roam freely, mingle, etc.), and to the negative rights to not be imprisoned (constrained), murdered, or cannibalized by humans, animal rightists are left with little alternative to veganism. Most people are willing to grant them that right. Confrontation arises, however, because animal rightists are not willing to grant people the right to be omnivores. The ideology of animal rights is akin to a fundamentalist religion that is intolerant of dissent or pluralism.

Because Judeo-Christian religion clashes with their beliefs, animal rightists might be expected to turn to secular humanist "religion." Those two belief systems don't mix well, however, because the word "humanism" is itself speciesism and hence taboo to animal rightists. Secular humanism has something to offer in its emphasis on humans and their perfectibility–if only "wicked" institutions such as animal producers and big corporate business would stop subverting the system.

That concept plus the ability to anchor ideology in "gut feelings" makes postmodernism, as outlined in chapter 2, a better fit for the animal rights mindset.

Several years ago one of my research assistants was a vegetarian who adhered to her diet with the legalism of an orthodox religious believer.[2] She did not eat meat because she believed that raising animals for slaughter was cruel. The philosophic question I posed to her was, "Is it better for an animal to have lived and been eaten than never to have lived at all?" The question is valid because the number of sentient animals likely would fall sharply if all people were vegetarians.

Even Singer equivocates on this issue. In the first edition of *Animal Liberation*, he rejected the argument that there would be less animal satisfaction because there would be fewer animals if people ate no meat. His grounds were that we cannot make judgments about the feelings of a nonexistent being. In his second edition, he says "But now I am not so sure" (p. 228).

According to a task force of ethicists and scientists (CAST 1997, p. 12), "Most animal rights activists call for the end of all food-animal production." Achieving such a draconian ban would require a sea change in culture along with strong, strictly enforced laws. The legal right granted by government requiring that animals to be treated like humans would clash with legal rights for humans to have pet animals; wear clothes (including shoes) made of leather, silk, fur, or wool; and to eat animal products. When rights clash, which or whose rights trump other rights?

A problem with animal rights or any moral rights based ethical system is that trade-offs become very difficult. To be sure, a modest rights agenda may be manageable. For example the Brambell Committee (Brambell 1965) to the British parliament call for a "five-freedom" bill of animal rights. The committee contended that an animal should at least be able to (1) turn around, (2) groom itself, (3) get up, (4) lie down, and (5) stretch its legs.

Europe has been a leader in government regulation of farm animal production systems to reduce animal cruelty. Sweden banned cages for laying hens, the United Kingdom banned crates for veal calves, and many countries and states banned cock fighting and dog fighting. We will return later to public policy for animals.

## VANDALISM AND VIOLENCE BY ANIMAL RIGHTISTS

On a moral crusade, the ends justify strong means. Fueled by deep-seated moral conviction that their crusade is righteous, animal rights advocates stop at almost nothing to further their cause. A consequence is harassment by animal rightists of people with other views. Animal rightists and radical environmentalists have been the most active terrorist groups in North America in recent years.

Bruce Friedrich (2001), People for the Ethical Treatment of Animals (PETA) campaign coordinator, provided this chilling call to action:

> These animals do have the same rights [as humans] to be free of pain and suffering at our hands. ...And considering the level of the atrocity and level of the suffering, I think it would be a great thing if, you know, all of these factories and slaughter houses and the laboratories and the banks that fund them exploded tomorrow. I think it is perfectly appropriate for people to take bricks and toss them through the windows and, you know,

everything else along the line. Alleluia to the people who are willing to do it.

PETA's coercive tactics have induced McDonald's, Burger King, and Wendys to use pork produced and processed under specified regulatory practices: unannounced slaughterhouse audits, no sow stalls, no tail docking, and no middle-teeth clipping ("Pork Alert" 2001, p. 2). In the case of laying hens, cages are allowed but birds must be given more room and must not be forced to molt.

In May 2002, PETA lifted its three-month-old ban on Safeway products and claimed credit for Safeway's announcement that it will conduct unannounced animal welfare inspections of its animal product suppliers. The retail grocer, one of the nation's largest, awaited guidelines for its animal care inspections from the Food Marketing Institute, a trade association representing food retailers ("Safeway Will Spot Check Seaboard Plants" 2002, p. 2).

The Animal Liberation Front (ALF) on its homepage says it "carries out direct action against animal abuse in the form of rescuing animals and causing financial loss to animal exploiters, usually, through damage and destruction of property" (ALF 2001, p. 3). The seemingly benign objective of "performing non-violent direct actions and liberations" (ALF 2001, p. 3) is in fact cruel. Many if not most animals released from confinement into the wild and not recovered will die a painful death because they are unprepared to survive in the wild. ALF's professed "non-violent direct actions and liberations" (p. 3) are especially violent against property. For many persons, property is an extension of themselves. Those who destroy property ultimately destroy people.

The "Report to Congress on the Extent and Effect of Domestic and International Terrorism on Animal Enterprises" listed 313 cases of violence involving animal rights activists in the 15 years preceding 1993. An example is Dr. Hyram Kitchess, dean of the veterinary school at the University of Tennessee, who was shot and killed on his farm in February 1990 (Wyant 1993, p. 26). Authorities suspected that the Animal Liberation Front or other animal rights group was responsible for the murder.

On the same day (September 11, 2001) that New York and Washington were attacked by terrorists, ALF claimed credit for firebombing a McDonald's outlet in Tucson, Arizona (Berman 2001, p. 1). No one was killed, but a bombing in France in 2000 killed a McDonald's employee.

Animal rightists were believed responsible for setting a fire and for leaving incendiary devices (unexploded) October 15, 2001, at a government facility for wild horses and burros in Nevada–a site where animal rightists committed arson a decade earlier. ALF claimed to have set fire to the Coulston Foundation primate research facility September 20 and to two meat trucks in New York in March 2001. ALF also claims credit for setting incendiary devices beneath trucks in Canada on Christmas Day, 2000 (Berman 2001, p. 2).

Graham Hall, a British journalist, was kidnapped at gunpoint in October 1999. The letters "ALF" four inches high were burned into his back with a branding iron. An ALF spokesperson commented that "people who make a living in this way have to expect from time to time to take the consequences of their actions" (Berman 2001, p. 2).

On May 6, 2002, a militant vegan enraged by Pim Fortuyn's willingness to legalize mink farming allegedly murdered Mr. Fortuyn, a major political party leader running for The Netherlands parliament and a potential force in the Dutch parliament. He was running strong in the parliament election before being killed ("The Political Legacy of Pim Fortuyn" 2002, p. 45).

The North American Animal Liberation Front has been especially active in vandalism to release animals raised for fur. Several incidents are reported in Iowa alone (Kompas, September 2000, p. 3B; October 2001, p. 1B):

- In August 1998, Isebrand Fur Farm in Jewel lost 3,000 mink and Hidden Valley Fur Farm in Guttenberg lost 330 foxes.
- In September 2000, 14,000 mink were released from the Earl Drewelow and Sons Fur Farm near New Hampton.
- In October 2001, vandals released 1,100 mink from the Scott Nelson farm near Ellsworth. Some 600 mink were rounded up, only to be released again by vandals a week later (Hupp 2001, p. 1B).

Several other animal vandalism cases took place in Iowa in recent years. The North American Animal Liberation Front took credit, but the vandals have not been caught.

Animal rightists claim such vandalism protects animals because it bankrupts farms. Such a claim has serious faults. One is that releasing farm animals raised in captivity is a death sentence because they cannot secure food in the wild. A second problem is that a bankrupt fur farm is replaced by another farm ready to meet the demand for fur. Finally, the financial and psychic destruction of fur farm operators and workers shows that the North American Animal Liberation Front values animals more than people.

Animal rights activists have tried to cut demand for fur by throwing paint and blood on women wearing fur coats. Demand for fur has not been seriously affected. As with other RA groups, law enforcement has not been effective. An injection of resolve and immediacy of the type apparent after the September 11, 2001, terrorist attack on the World Trade Towers is overdue for law enforcement to address RA violence.

In summary, animal rightists like everyone else are entitled to express their views and lead public protests in order to change laws and public opinion. But vandalism and violence against animals and people to help animals not only violate the law, they also violate simple standards of common decency.

## PUBLIC POLICY

Public policy decisions depend in part on what kind of life farm animals lead. Even vegetarians might be willing to tolerate human use of animals if they are treated well and live a life of contentment. A problem is that despite the useful information provided by animal psychologists and physiologists, human judgments of animal satisfaction are not very objective.

Modern, industrial farming methods do not necessarily mean less-humane treatment of animals. When I was growing up on a "preindustrial" mixed crop-livestock farm in Iowa prior to World War II, we treated our animals pretty much like other farmers treated their animals. In winter we kept our dairy cows in stanchions except to let them run in the barnyard for a short recess each day to drink from the outdoor water tank. Today, large dairy operations use free-stall housing so dairy cows can more freely move around. Drinking water is readily available indoors. Animals are watched carefully for signs of illness. When warranted, treatment is immediate and thorough.

On the farm, we slaughtered animals for home consumption. Animals were not stunned before slaughter. My father inserted a large butcher knife into the hog's jugular vein in its chest. It bled to death. Today, hogs and other animals are stunned before slaughter.

On our farm like on other farms, animals were allowed to roam in the summer (good), but were subject to the indignities inflicted by wild animals and the elements. Thus, treatment of animals has changed greatly in recent decades, sometimes for better and sometimes for worse.

In the United States the Food and Drug Administration, the Animal and Plant Health Inspection Service, the Environmental Protection Agency, and numerous other federal and state agencies work to ensure that food is safe. The Animal Welfare Act, the Animal Transport Act, the Humane Slaughter Act, and various other laws and agencies work to protect animals from cruelty by humans. (Cruelty by animals to other animals is less often prohibited.)

Scientific use of animals is regulated by the Animal Welfare Act of 1966 as amended in 1970 and 1976 and by the Improved Standards for Laboratory Animals Act of 1985 (CAST 1997, p. 14). These regulations are enforced by the U.S. Department of Agriculture, but of interest is that the laws specifically exclude farm animals. Several national producer associations including those for veal, milk, pork, and turkey have established voluntary guidelines for sound animal care (CAST 1997, p. 15).

The Humane Slaughter Act states that "the slaughtering of livestock and the handling of livestock in connection with slaughter shall be carried out only by humane methods." People for the Ethical Treatment of Animals (PETA), an organization frequently taking an animal rightist position, in late 2001 filed a petition with the U.S. Department of Agriculture contending that the agency needs to apply the act to animals handled and killed on factory farms. Animal rights advocates are especially critical of "factory farms" with their battery cages for laying hens, sow stalls, and crowding of animals inevitably confined indoors.

Castration, dehorning, ear notching, branding, vaccination, and tail docking are traumatic. Considerable progress has been and more needs to be made in devising and using means such as anesthetics to reduce that trauma. Many practices such as "milk fed" veal that Singer (1990, p. 129) regarded as the "most morally repugnant" of farming practices are now rare. In that system veal calves were kept in dark pens too small to turn around in and fed a diet low in iron to keep flesh white and tender before slaughter.

My unscientific appraisal based on spending many years around a number of species of farm animals is that for the most part the animal's life is one of contentment if not satisfaction. My conclusion that an animal is better to have lived and been eaten than never to have lived at all does not diminish the case for treating animals humanely (for elaboration of issues, see Rowan 1993).

**The Question of Diet?**

In affluent nations, chronic excessive calorie intake is the most common form of malnutrition. Over half of American adults are overweight and a sizable share are obese, in part because chronic excessive calorie intake is made attractive and easy with "fast foods" containing high proportions of fat in meat and cheese. Deaths attributed to chronic excess calorie intake total some 300,000 per year in America, the second largest source (after smoking) of preventable mortality. A vegetarian or vegan diet can reduce obesity and be as healthy as other diets.

Animal products have a place in a tasty, healthy diet, as apparent from the well-known food pyramid published by the nutrition people in the U.S. Department of Agriculture. Lean meat, nonfat milk, and eggs are excellent sources of protein and are low in fat. Obtaining all essential nutrients is a challenge for vegetarians and is even more challenging for vegans. Vegetarians need a wide variety of plant foods to obtain all essential amino acids. Vitamin $B_{12}$ is an essential nutrient that is difficult to find in plant foods, although soy sauce contains $B_{12}$. Countries that rely mostly on vegetarian diets have many stunted people, in part because people are poor and uneducated in nutrition. Malnutrition is widespread. As income rises, people can afford to, prefer to, and do shift their diets to include more animal products.

The concluding point here is that a sound diet may or may not include animal products, but it does demand discipline and common sense, especially to avoid chronic overeating in affluent societies. To reiterate, the policy quarrel of conventional agriculturalists with animal rightists is not about vegetarian or vegan diets. Rather, the quarrel is over tactics used by animal rightists to achieve their objectives. The appropriate response to the ubiquitous problem of excessive calorie intake is not violence but rather is better nutritional education and food technology.

**Subtherapeutic Antibiotics**

Subtherapeutic use of antibiotics is a contentious policy issue not directly affecting animal rights, but affecting the welfare of animals and people. Animals given subtherapeutic antibiotics suffer less illness and mortality, but are likely to be raised in a large, confinement operation. The latter condition arises because antibiotics have helped to make confinement operations possible by avoiding catastrophic death losses that can occur as large numbers of animals are kept in close proximity to each other.

An example is enrofloxacin, an antibiotic similar to the wide-spectrum, powerful Cipro used to treat people with anthrax. Enrofloxacin protects chickens from respiratory diseases, but does not quite eliminate a different strain of bacteria, *campylobacter,* in the chicken's digestive system (Gorman 2002, p. 100). Some enrofloxacin and Cipro-resistant bacteria leave the chicken to eventually find their way into *campylobacter* that causes food poisoning in humans. Before 1996, the number of *campylobactor* infections resistant to Cipro in humans was negligible. By 1999, the resistant proportion had grown to 18 percent (Gorman, p. 101). The portion of that resistance attributed to farm use of antibiotics is not known, but will be studied in greater depth in the future.

Federal regulation of antibiotic use for disease prevention and growth stimulation may be forthcoming. In 2002, Senator Edward Kennedy introduced a bill in the U.S. Senate to phase out "routine feeding of medically important antibiotics to healthy farm animals." A similar bill was proposed in the U.S. House. Both bills had considerable support and endorsements, including the American Medical Association. The impact of implementing the legislation on agriculture depends on the availability of "less medically important" antibiotics that can be used on "healthy" farm animals.

Even before action by Congress, signs of industry self-regulation were promising. Burros (2002, p. 1) reported that "three companies–Tyson Foods, Perdue Farms, and Foster Farms, producing a third of the chicken consumed by Americans each year–say they have voluntarily taken antibiotics out of feed for healthy chickens." Information is not available from other poultry producers, but they are believed to be following patterns set by the industry leaders. In addition, Burros reports that the industry is turning away from use of Cipro-related antibiotics for any purpose (p. 1).

A task force of scientists on the Well-Being of Farm Animals (CAST 1997) has specified an ambitious set of research requirements to devise a more scientifically based set of guidelines and regulations for animal treatment. A major multidisciplinary effort is called for to address animal welfare from the standpoint of bioethics, pathology, physiology, and other disciplines.

**Dealing with CAFOs**

Residents of local communities, states, and the nation are ambivalent about whether farm structure regulations should seek to (1) preserve small farms and discourage large farms because "small is beautiful" and "big is bad"; (2) internalize farming externalities such as flies, odors, and water pollution; or (3) seek local ownership and control of farms. Policy prescriptions differ depending on the goal. If the goal is control of flies and odor, there is no reason to confine regulations to large farms. If the goal is local ownership and control, then large, efficient, locally owned farms might be economically viable and preferred over marginal small farms. If the goal is to have many small farms, then local nonfarm job creation is a useful strategy to supplement the low income of small farm operators.

Concentrated (sometimes referred to as confined) animal feeding operations called CAFOs are anathema to animal rights activists. These operations are the subject of much policy debate, some of which is reviewed here. A CAFO contains 1,000 animal units or more. That size is 700 dairy cows, 1,000 beef cows, 2,500 hogs, or 100,000 laying hens. Many states require such facilities to formulate and follow operating permits meeting Environmental Protection Agency regulations.

Large-scale animal feeding operations are often called *factory farms*. That name is intended as a pejorative when used by animal rightists, but is descriptive. I have toured large laying-hen operations where semitrailer trucks unload feed at one end of the building while semitrailer trucks at the other end of the building haul away cartons of eggs untouched by humans.

Are such operations good or bad? CAFOs produce more efficiently than smaller farms (Tweeten and Flora 2001). Lower costs of production are passed to consumers–a special benefit to low-income consumers. A state or locality that attracts or "grows" CAFOs increases employment and income. Blue-collar workers are especially likely to find employment opportunities they would not have on small family farms. Real estate value may fall near the facility, but is likely to increase in the greater community because the local community and state economic base is strengthened. Input supply such as veterinarian services and farm commodity processing facilities such as packing plants made feasible by CAFOs provide local jobs and may provide services to sustain local small family farms. Local farmers find local markets for their crops at prices somewhat higher than in the absence of CAFOs.

Although CAFOs create jobs locally, they displace workers nationally because they use resources more efficiently. The Buckeye Egg farm in Ohio, for example, with 15 million laying hens replaces a large number of small family farm laying operations of the 300-hen size I grew up on.

Critics accuse CAFOs of abusing the environment and animal rights. A central issue is whether CAFOs would be more economically efficient than smaller farms and hence would prosper if all farms where required to internalize externalities–properly controlling flies, odors, and waste disposal. The answer to that question is elusive. Experts on farm structure such as Martin and Zering (1997, p. 20) and Boehlje et al. (1996) contend that costs per unit of output would continue to be lower on large farms than on small farms if externalities were internalized.

A reason is that efficient waste disposal is characterized by economies of size. By spreading high equipment (overhead) costs of waste disposal over many units of production, larger operations can have lower per unit costs to convert animal waste into methane gas for fuel, and ash or compost for low-cost storage, transport, and injection into soil for fertilizer.

Several policy conclusions follow:

- The core trade-off is whether the nation wants lower food costs and larger livestock operations or higher food costs and small family farm livestock operations. The political process rather than social scientists must resolve that dilemma.
- Local areas do not face a choice of small versus large poultry, dairy, hog, or cattle feeding operations. Rather, the choice is between large operations or no operations because small operations cannot compete unless

subsidized by off-farm income. Local areas or states that attempt to protect their small farms by baring large, corporate, or vertically integrated operations will find such operations going to other areas or states. To be sure, small livestock farms have higher local employment and purchase multipliers than the more efficient large farms, but such multipliers are meaningless when small farms lack efficiency required to survive.

- Only national legislation has a chance to halt large operations. In the unlikely event that such legislation is passed, it is unlikely to be effective. Farmers are very clever at circumventing costly bureaucratic regulations. Also, large operations would move to Canada or Mexico to produce for export to the United States.

These considerations need not preclude imposition of reasonable environmental and animal welfare regulations. It is well to remember, however, that such regulations, if imposed fairly on all farms, are likely to disadvantage small farms as much as large farms. One reason is because the environmental and animal welfare practices that animal rightists find objectionable on large farms are mostly the result of technology and management rather than size, corporate ownership, or vertical integration. That is, small farms like large farms need modern technology to deal with problems of waste disposal and animal welfare. Corporate ownership and production contracts provide a quick, convenient path to achieve size economies that today's small operations have incentives to grow into with time.

Many of the environmental and animal welfare problems of large farms are, in fact, the result of poor management that as least as often troubles small farms. Problems of externalities, poor management, and inhumane (by some definition) treatment of animals would remain and might intensify if CAFOs were outlawed. One reason is because controls are more easily administered to a few large farms than to many small farms.

**A Utilitarian System**

Although Singer seems squarely in the animal rights camp with his statement that "all animals are equal," some classify him as a utilitarian who considers benefits and costs in deciding how animals can be used. For example, he did not rule out use of animals for some purposes such as medical use where human lives might be saved. Such equivocation signals the ethical complexity of animal use.

Utilitarian systems that allow balancing of benefits relative to costs of various forms of animal care and use have much appeal. The price system coupled with brand names and labeling allows people individually to make decisions in the market regarding use of animals based on their unique perceptions of benefits relative to costs. Costs often are higher for animal production systems providing the greatest animal rights. Eggs produced in caged housing are estimated to cost only half as much to produce as eggs produced by free-range chickens (CAST 1997). A consumer who highly values the freedom of laying hens to roam freely presumably would consider benefits of free-range eggs to exceed their cost and would buy them in the supermarkets. Consumers who value freedom of laying hens less would purchase the cheaper cage-produced eggs.

Voluntary labels can identify processes used to produce foods. The government allows a "Free Farmed" seal on products from farms that eliminate cages for laying hens and forced molting, and withdrawal of food and water to increase egg production. Dairy products can be labeled as coming from cows having access to pasture. Standards also are being developed for hog farms. The Farm Animal Services agency, set up by the American Humane Association, inspects farms that want to use the Farm Animal Services' label. The U.S. Department of Agriculture monitors the inspections.

The utilitarian view that individual persons can best chose in the marketplace whether to consume meat or not has more appeal if farm animals are provided a favorable quality of life. Another argument against human use of animals is that food derived from animals requires more resources and hence is more damaging to the environment than food derived directly from plants. The counter argument is that the world has plenty of resources to feed all people an adequate or preferred diet. Furthermore, livestock frequently utilize grazing land unsuited for cropland or other purposes.

## CONCLUSIONS

A sense of community, a form of social capital, is the "glue" that binds a nation together. In the case of animal agriculture, the glue has lost some stickiness.

Some years ago I pulled into an Interstate Highway 35 rest stop near a small Iowa town, Lamoni, where a plaque listed the utopian societies that once existed in the area. Two things impressed me: the many organizations on the list, and that I had never heard of any of them. The masses did not embrace the "compelling" ideologies that had motivated the founders, and the organizations faded into history.

With breathtaking audacity, animal rights ideologues are mounting a frontal assault on Western culture. Activists are engaged in a variety of means including vandalism to achieve their ends. The contest for the hearts and minds of people is unlikely to be won by violence, however. Dialogue and use of established political institutions are critical. So is respect for commandments to not bear false witness, steal from, or kill our neighbor. The dialogue is likely to continue over whether animals are our neighbor. Will animal rights organizations become another entry on a plaque somewhere listing failed utopian organizations?

## NOTES

1. Singer (1990, p. viii) recognized that he was by no means the first to condemn existing treatment of animals and to speak for animal rights. Henry Salt wrote *Animal Rights* in 1892, but it mostly gathered dust in the British Museum. Another pioneer was Ruth Harrison whose book *Animal Machines* published in 1964 breathed life into the animal rights movement, especially in Britain.
2. Initially, my vegetarian research assistant would not eat cheese because it was made using rennin taken from stomachs of slaughtered calves. When I informed her that an enzyme produced by bioengineered *E. coli* had replaced such rennin, she was thrilled to return to eating the cheese pizza she loved.

## REFERENCES

"Absolutely Nuts About Numbers?" *Desert Sun*, January 24, 2002, p. A6.

ALF (Animal Liberation Front). http://www.hedweb.com/alffaq.htm. December 12, 2001.

Berman, R. "Enemies Here Threaten Food." http://www.mannyculture.com/oped_detail.cfm?OPED_ID=133. December 13, 2001.

Boehlje, M. "Industrialization of Agriculture: What Are the Choices?" *Choices*, First quarter 1996, pp. 30-33.

Brambell, F. W. R. *Report of Technical Committee to Enquire into the Welfare of Animals Kept under Intensive Husbandry Systems*. Command Paper 2836. London, UK: HM Stationery Office, 1965.

Burros, M. "Poultry Industry Quietly Cuts Back on Antibiotic Use." *New York Times,* February 10, 2002, p. 1.

CAST (Council for Agricultural Science and Technology). *The Well-Being of Agricultural Animals*. Task Force Report No. 130. Ames: CAST, September 1997.

D'Amico, T. "Charting a Course in the New Millennium." http://articles.animalconcerns.org/ar-voices/teresa/ar2000.html. December 13, 2001.

Friedrich, B. ( PETA's Vegan Campaign coordinator). Speech to Animal Rights 2001 convention. http://www.mannyculture.com, 2001.

Gorman, C. "Playing Chicken with Our Antibiotics." *Time,* January 22, 2002, pp. 100, 101.

Harrison, R. *Animal Machines: The New Factory Farming Industry*. London, UK: Vincent Stuart Publishers, Ltd., 1964.

Hupp, S. "Mink Spring Again; Farm Near Failure." *Des Moines Register*, October 24, 2001, p. 1B.

Kompas, K. "Arrests Unlikely in Mink Case." *Des Moines Register*, September 9, 2000, p. 3B

Kompas, K. "Vandals Release 1,400 Mink." *Des Moines Register*, October 18, 2000, p. 1B.

Martin, L., and K. Zering. *Relationships between Industrialized Agriculture and Environmental Consequences: The Case of Vertical Coordination in Broilers and Hogs*. Staff Paper 97-6. East Lansing: Department of Agricultural Economics, Michigan State University, 1997.

"The Political Legacy of Pim Fortuyn." *The Economist,* May 11, 2002, pp. 45, 46.

"Pork Alert." Vol. 2, issue 45. PorkAlert@vancepublishing.com. November 27, 2001.

Rowan, A. "Animal Well-Being: Some Key Philosophical, Ethical, Political, and Public Issues Affecting Food Animal Agriculture." In *Food Animal Well-Being,* pp. 23-36. West Lafayette, IN: Office of Agricultural Research Programs, Purdue University, 1993.

"Safeway Will Spot Check Safeway Plants." *Pork Alert*. Pork_Alert.ue.516176.t18458770@ixs1.net, May 21, 2002.

Salt, H. *Animal Rights*. London, 1892. Reprint, Clarks Summit, PA: International Society for Animal Rights, 1980.

Singer, Peter. *Animal Liberation*. 2d ed.. New York: Random House. 1990.

Tweeten, L., and C. Flora. *Vertical Coordination of Agriculture in Farming-Dependent Counties*. Task Force Report No. 137. Ames, IA: Council for Agricultural Science and Technology, March 2001.

Wyant, S. "Animal Terrorists Out of Hand." *Ohio Farmer*, November 1993, p. 26.

# 7

# Agrarian Populism and Farm Fundamentalism

## INTRODUCTION

This is the first of three chapters dealing with agrarian populism. Populism here refers to thinking that offers appealing, simple, and straightforward solutions to complex problems–solutions that turn out to be wrong.

The market alone offers little to populists. Populists seek solutions to problems in the political arena. Those solutions can be costly because they often attract numerous dedicated supporters who convince lawmakers to redistribute wealth to them.

Agrarian populism applies to rural communities and agriculture, but here mostly is about the farm sector. Farm fundamentalism, a major belief system supporting agrarian populism, is discussed at length in this chapter and is contrasted with democratic capitalism, the dominant American belief system. Chapter 8 is about populist myths. Chapter 9 relates populism to farm organizations and protest movements.

Radical agriculturists (RAs) and populist agriculturalists (PAs) have much in common. Each has postmodernist tendencies, believing that powerful agribusiness interests have captured and perverted sound science and economics. In their view, scientists and economists have "sold out" to serve the interests of agribusiness. Like RAs, PAs distrust or downright dislike "middlemen," "big business," corporate conglomerates, and profit. PAs and RAs find common cause in their distrust of the agricultural establishment. Hence, in the postmodernist tradition, PAs and RAs must rely on their personal insights and gut feelings to augment if not replace conventional thinking.

However, unlike radical agriculturalists who are mostly nonfarmers interested in the problems of agriculture and the food system, PAs are mostly farmers and farm organizations. They may be on the fringe and ordinarily do not control the agenda of the agricultural establishment, but they nonetheless are part of it. Unlike radical agriculturalists whose mission is often idealistic in ostensibly serving others (consumers, small farmers, and the environment), the PA mission is self-serving. Above all, farm populists want to make more money, and they invoke precepts of farm fundamentalism to serve their objective.

Populism has mostly been a blight on the American economy. Agrarian populists have claimed credit for farm commodity program, minimum wage, antitrust, and interstate commerce legislation. Each of these has shortcomings. Much interstate commerce legislation painstakingly assembled by populists in the nineteenth century has been dismantled through deregulation in recent years to the benefit of consumers and the nation's economy. Minimum wage legislation continues to create unemployment among the most disadvantaged would-be workers–those who contribute less to the value of a prospective employer's output than the minimum wage. (A wage supplement, as I have outlined elsewhere, coupled with the current earned income tax credit, offers more for disadvantaged workers as well as for the national economy. See Tweeten 1989, pp. 308-13.) Worthy antitrust legislation has been put in place, but would likely have emerged even without muckraking populist campaigns.

General condemnation of agribusiness (firms supplying farm inputs and marketing farm products) is a shabby legacy of populism. Rising productivity of American agriculture contributing mightily to this nation's high living standards rests on three pillars: agribusiness, science, and production agriculture. These pillars can be likened to a three-legged stool. Anyone who has sat on a three-legged stool while milking cows can appreciate how removal of one of those legs collapses the stool. Performance of agribusiness persons, like that of scientists and farmers, has been exemplary and the envy of the world. Without that agribusiness "leg," food costs would be much higher, hunger more widespread, and living standards much lower. Farm as well as nonfarm households would have lower income and wealth. In contrast to the situation in the 1996-2001 period, farm households likely would have lower income than that of nonfarm households in the absence of innovative and efficient agribusiness.

Antitrust laws are in place and appropriately have been used to curb predatory, exclusionary, and price fixing behavior of agribusiness firms. But the accusation by populists that agribusinesses willfully destroy family farms and cause farm prices to be lower than would be the case with a more competitive agribusiness sector is not borne out by analysis (Persaud and Tweeten 2002).

That and other myths perpetrated by populists are scrutinized in chapter 8. These myths and farm fundamentalist attitudes held by farmers and nonfarmers alike help to explain why farm commodity programs benefiting a relatively wealthy 0.2 percent of the nation's population are not vetoed by the 99.8 percent of Americans who are paying for the programs.

## FARM FUNDAMENTALISM AND DEMOCRATIC CAPITALISM

Two reasonable persons can look at the same facts and reach opposite policy conclusions. A major reason is because their goals, values, and beliefs differ. It has been said that economists can provide a half-dozen ways to solve the major farm problems, but progress toward a solution rests on resolution of conflicts in goals, values, and beliefs (Heady 1961, p. vi).

*Goals* are ends or objectives toward which behavior is directed. *Beliefs* whether correct or incorrect are what we hold to be true–the perceived nature of reality. *Values*–feelings of what is desirable or ought to be–are the standards of preference that guide behavior including political behavior. Goals, values, and beliefs may be based on fact and reason, but they are often subjective and intuitive feelings deeply rooted in the psyche from accumulated experiences, from culture, from genetic disposition, and from ramblings of long-departed philosophers. Whatever the source, values and beliefs often overshadow economic facts and careful analysis in policy-making. We cannot understand farm and food politics and policy unless we understand values and beliefs.

America is of two minds about agriculture. One submits to the creed of *farm fundamentalism,* emphasizing belief in the primacy of agriculture and the family farm. The other submits to the creed of *democratic capitalism,* emphasizing belief in the primacy of free enterprise and authority of the people at large expressed through the political system.

## FARM FUNDAMENTALISM

Articles in the farm fundamentalist creed include:

1. Agriculture is the most basic occupation in our society, and all other occupations depend on it.
2. Agriculture must prosper if the nation is to prosper.
3. Farmers are better citizens, have higher morals, and are more committed to traditional American values than are other people; indeed the nation's moral and social character depends on farmers.
4. The family farm must be preserved because it is a vital part of our heritage.
5. Farming is a way of life, a more satisfying occupation than others.
6. The person who tills it should own the land.
7. Anyone who wants to farm should be free to do so (Paarlberg 1964, p. 3).
8. A farmer should be his or her own boss–independent, rugged, and self-reliant.
9. The family farm is the ideal nuclear family unit where the family works, plays, prays, and, in general, lives together for mutual harmony and support.

### Origins of Farm Fundamentalism

Farm fundamentalism has origins in the bucolic images of the Bible. God's Chosen People were pastoral and agricultural. The image was of the *Good* Shepherd. Part of the Judeo-Christian tradition was the natural law doctrine of the Catholic Church that viewed the family farm as the principal vehicle for social order, progress, and reproduction. Pope Pius XII stated six decades ago that "only that stability rooted in one's holding [farm] makes of the family farm the vital and most perfect and fecund cell of society joining up in a brilliant manner in progressive cohesion the present and future generations" (Pius 1943). Pius viewed urbanization and industrialization as a threat, noting with alarm in 1951 that "Today it can be said that the destiny of all mankind is at stake. Will men be successful or not in balancing this influence [of urban-industrial society] in such a way as to preserve...the rural world?" (Pius 1951).

American agricultural fundamentalism stems from a truism: when farming was the major source of the nation's wealth, most of the people lived on farms. The small family farm was the most efficient economic unit. But agriculture fundamentalism became much more. That fundamentalism holds that farming is a divine calling in which God and humans walk hand in hand to supply the physical needs of all people(Fite 1962, p. 1203).

Agrarian fundamentalism's most articulate and compelling progenitor in America was Thomas Jefferson. He stated:

> Those who labor in the earth are the chosen people of God, if ever he had a chosen people.... Corruption of morals in the mass of cultivators is a phenomenon of which no age nor nation has furnished an example. It is the mark set on those who, not looking up to heaven to their own toil and industry as do husbandmen for their subsistence, depend for it on the casualties and caprice of customers. Dependence begets subservience and venality, suffocates the germ of virtue, and prepares fit tools for the designs of ambition. (1788, p. 175)

"Cultivators of the earth," Jefferson (1926, p. 15) wrote to John Jay in 1785, "are the most valuable citizens. They are the most vigorous, and they are tied to their country and wedded to its liberty and interests by the most lasting bonds."

The fundamentalist theme again is apparent in William Jennings Bryan's "Cross of Gold" speech: "Burn down your cities and leave your farms, and your cities will spring up again as if by magic; but destroy our farms and the grass will grow in the streets of every city in the country."

Some academics have joined in viewing the family farm as the wellspring of virtue. The family farm according to Harvard sociologist Carle Zimmerman (1950) was the form of agriculture "in which home, community, business, land, and domestic family are institutionalized into a living unit which seeks to perpetuate itself over many generations." The family farm feeds society with virtue "as the uplands feed the streams and the streams in turn the broad rivers of life" (Zimmerman 1950).

At an international conference on farm policy, Horace Hamilton (1946, pp. 100-113) said the family farm produced "men of strong character and moral consciousness." By way of contrast, the farm-labor camps found on large-scale commercial farms spurred "pool rooms, honky-tonks, cheap picture shows..., and flashy, back-slapping personalities." Based on a California study, Goldschmidt (1946) contended that a community surrounded by family farms was more economically and socially viable than a community surrounded by large corporate-industrial farms.

Continuity in farm fundamentalism to more recent times is not hard to find. Catholic Father G. Speltz (1963, p. 46) stated that, "the rural values such as reverence for the soil, love of God, love of fatherland, and willing acceptance of honest toil are fundamentally spiritual" and went on to recognize farmers as the source of such values, noting that, "even if this nation could dispense with most of its farmers, there would be the question of whether it could remain strong without the type of man agriculture produces." Maurice Dingman (1986), Catholic Bishop of Des Moines, noting the virtues of the family farm, lamented:

> The laissez faire approach has allowed the harsh forces of uncontrolled competition to drive less prosperous farmers out of agriculture. The adaptive approach goes so far as to employ the power influence of government and educational institutions, including land-grant universities, to accelerate the migration of farm families from the land. This should not have been permitted. That policy has been immoral, unethical, unjust, disastrous, motivated by greed, destructive, leading inevitably to conditions similar to Central America.

**Public Perceptions of Agricultural Fundamentalism**

Farm fundamentalism is very much alive. It remains a unifying symbol, a rich legacy of tradition cherished by the public. There is evidence that people think of farming not in terms of fact but in terms of deeply felt beliefs and values of what the nation ought to be. Public opinion polls have probed in some depth the public perception of agricultural issues. One was the AgFocus/Gallup survey of 1,507 persons by telephone interview in 1985 (National Issues Forum 1987). Another was a nationwide mail survey of 3,239 adult Americans conducted in the spring of 1986 by the S-198 project on socioeconomic dimensions of technological change, natural resource use, and agricultural structure (see Jordan and Tweeten 1987).

The AgFocus study (National Issues Forum 1986) concluded that

> people's attitudes and views about agriculture are a product, not so much of their specific knowledge about agriculture, but of their overall personal and political philosophies. Few Americans appear to have brought their knowledge about agriculture to bear on their opinions of it. (p. 3)

A surprisingly high proportion, over half of the S-198 survey respondents chosen at random from the nation's adult population, had been exposed to farming through relatives or friends who were farmers. Yet, respondents' knowledge of agriculture tended to be low, varying among the groups and issues being considered. Where perceptions were incorrect, they tended to favor rather than work against the position of farmers. The implication is that better education of the public at large will not necessarily translate into greater support for farmers in the political arena.

Three-fourths of the respondents in the S-198 study agreed (incorrectly) that "today most farmers are in financial trouble" (Jordan and Tweeten 1987). Four-fifths of the respondents agreed that "agriculture is the most basic occupation in our society, and almost all other occupations depend on it." A slightly higher percentage (82) agreed that the family farm must be preserved because it is a vital part of our heritage.

Farm fundamentalism was strongest among respondents who are women, white, aged, married, less-educated, low-income, conservative, religious, and close to agriculture (Tweeten 1989, p. 77). Thus, support for farm fundamentalism may diminish as the nation becomes more urban, affluent, educated, distant from agriculture, and as the young age (unless they change beliefs with age). The surprise was not that fundamentalism statistically differed among categories for all but political party affiliation, nor that 99 percent of people on farms subscribed to farm fundamentalism. The surprise was that a large majority of people in all categories including two-thirds of those living in large cities held farm fundamentalist beliefs.

Such beliefs do not necessarily translate into favoring special treatment of farmers. Only 24 percent of respondents agreed that most consumers would be willing to have food prices raised to preserve the family farm. A 1985 poll of consumers found that 68 percent of respondents would be willing to pay an extra 1 percent of their grocery bill to preserve America's family farms (National Issues Forum 1986, p. 39). The image of farming given by respondents was one of a preferred way of life but characterized by hard physical labor, high risk, and low economic returns. Other careers were more favored by respondents for themselves and for their children.

Respondents simultaneously subscribed to farm fundamentalism and democratic capitalism. The S-198 study found that 63 percent of the respondents agreed and only 33 percent disagreed with the statement that "the government should treat farms just like other businesses." Also more respondents agreed than disagreed with the statement that "farmers should compete in a free market without government support." However, other parts of the survey revealed strong support for Treasury transfers to farmers.

Are Americans taking agricultural policy positions based on facts or feelings? Unfortunately, that question cannot be answered with reliability, and further study is warranted. Fragmented information suggests that the public is poorly informed regarding the agricultural economy. A survey of consumers in 1985 found that 46 percent of respondents thought that food raised on a corporate farm would be more expensive than food raised on a family farm (National Issues Forum 1986, p. 30).

In reality, no evidence indicates that food produced on corporate farms would be any more expensive or of lower quality than food produced on family farms. Corporate business organization itself has little direct impact on cost of production or food quality. Indirect impacts are apparent, however. Corporate business organization offers advantages in handling risk and acquiring assets to form a large farm necessary to achieve economies of size known to permeate agricultural production (Tweeten 1989, pp. 123-25).

### Farm Fundamentalism under Scrutiny

Many of the beliefs that constitute the creed of farm fundamentalism do not stand scrutiny. Each of the articles in the fundamentalist creed is examined briefly below in the order in which they appeared earlier in this chapter:

1. Farming *is* a basic occupation but it is one among many and is not necessarily *the* most basic. Provision of water is more basic to life than food, but society does not award the water utility industry or its operators special status. The economic value of an industry is not determined by whether it is natural resource based, whether many people and industries use its output, or whether it is basic to sustaining life. Rather, its value is determined by its contribution *at the margin* to satisfactions of society. Low farm and food prices, surpluses, and production controls are signals that society places little value on additional farm output. A person drowning in water places a negative value on having more water although some water is essential for life. A society "drowning" in farm commodity production places negative value on additional output.
2. Agriculture is such a small part of the national economy (less than 2 percent of Gross National Product) that the industry can be depressed while the nation's economy prospers–as in the 1920s or mid-1980s. Similarly, agriculture can prosper while the nation is depressed as in 1974-75 and 1990-91.
3. It is not possible to objectively state that farm people are superior to others. Farmers have lost much of their uniqueness. To be sure, farmers continue to have lower crime and divorce rates and to have higher churchgoing and birth rates than others (Drury and Tweeten 1997). On the other hand, farmers seem to be less committed than others to virtues of racial tolerance, sexual equality, and concern for the disadvantaged (Burchinal 1964, pp. 160-62, 168-70). The data need updating, so at best we can only conclude that farmers are neither unequivocally better nor worse than others. But it is possible to say that farmers' actions, goals, values, and beliefs are becoming more like those of nonfarmers. A related issue, as Carlson (1986, p. 20) put it, is that "America's farms increasingly resemble old-age homes. We have exhausted their human capital. We must recognize that our nation's family farmers can no longer serve as our peasant class, reinvigorating an otherwise failing social order and providing a stream of surplus youth for factories and offices. Put simply, the numbers are no longer there."
4. Whether to preserve the family farm is an *option,* not a "must." The decision to preserve or not to preserve is for the political process and not for social scientists to make. Social scientists inform; the political system makes decisions whether to intervene in the market to preserve the family farm. If the political process is representative, social scientists have no basis to object if an informed society valuing the family farm for its intrinsic worth is willing to and does make necessary sacrifices in taxes and food costs to preserve family farms. The quarrel that many economists have with populists is not whether to preserve family farms, but rather is the "bait and switch" tactic of populists. They justify spending billions of taxpayers' dollars in the name of preserving family farms but in fact spend the money in ways that destroy family farms. They call for federal subsidies to save small farms, but make certain that most of the benefits go to commercial farms.
5. Surveys of farm and nonfarm residents reveal preferences for a farm way of life but stressful economic conditions give rise to ambivalence apparent in

this song sung by farm fundamentalist activists at a protest rally:

Now some folks say
There ain't no hell
But they don't farm
So they can't tell.

In a 1986 article in *The Witness,* a religious newspaper in Iowa supportive of farm fundamentalism, as punishment for an alleged transgression I was sentenced to spend purgatory on a 300-acre Iowa farm–another indication of the ambivalence that family farmers and others feel toward the family farm. Such feelings prompted Jeffrey Pasley (1986, p. 27) to write, "Given the conditions of life on the family farm, if ITT or Chevron or Tenneco really do [sic] try to force some family farmers off their land, they might well be doing them a favor."

Sociological studies (see Coughenour and Tweeten 1986, for review) show that farmers are more satisfied than persons in other occupations with working conditions (farm way of life), but on the whole are no more happy or satisfied than others because they feel poorly rewarded for their work. This finding is not a surprise–most farmers are on units too small to be economically efficient or to provide an adequate income for a full-time farm operator and family. Taxpayers have intervened to generously assist farmers, but such assistance is unusual for sectors or hobbies–taxpayers ordinarily do not subsidize people just because they enjoy their hobby or their work. Farmers view economic instability as a hardship and they seek relief from taxpayers. Speculators in the futures and stock markets also face great uncertainty but the public does not bail out losers in the Chicago Board of Trade or New York Stock Exchange. President Reagan's budget director David Stockman raised a storm of controversy and was chastised by the president for commenting, "I can't see why taxpayers should refinance bad debt [of farmers] that was willingly incurred by consenting adults who…thought they could get rich."

6. Perhaps the operator who tills it should own a farm, but an economic farming unit of sufficient size to be efficient and provide an adequate family living now requires $2 million or more of assets. Few farming entrants can be owner-operators of such a unit without massive assistance from parents or other outside sources. To get started or expand in commercial farming, most operators find it advantageous to reduce cash-flow pressures by renting at least some land.
7. Massive subsidies would be required to allow everyone to farm who wants to farm regardless of their managerial or financial capabilities. The federal government and several state governments provide concessional lending to help more persons get started in farming. Such generosity lacks merit for an industry allegedly troubled by excess capacity to produce and by more farmers than can make a living. In reality, most large farms would do well financially and most small farms would successfully support their hobby farming with off-farm income in the absence of government commodity programs. Because

current commodity programs have created an insidious economic dependency by farmers on taxpayers, a transition "detox" program would be necessary to cut down on farm financial hardship while easing farmers back into markets.
8. It is nice to be one's own boss, but increasingly a farm operator must share decisions with spouse, the banker, and the government. Making agriculture a public utility with commodity prices set to provide an assured rate of return clashes with the efficiency tradition of democratic capitalism discussed later.
9. Farm families follow lifestyles pretty much like those of their nonfarm neighbors. Most farm operators or their spouse work off the farm. Kids are in extracurricular after-school activities, and carpooling is widespread. Some spouses of operators work because they have no choice financially; others work because they enjoy the discretionary spending money and the social interaction. In short, farm people are good people; so are plumbers, nurses, factory workers, and their families.

## DEMOCRATIC CAPITALISM

A second creed or set of beliefs, democratic capitalism, is also prominent in farm and nonfarm America. One cannot understand farm policy without understanding it as well as farm fundamentalism. Although the two creeds sharply conflict, both are widely held, *often by the same person!* Farm fundamentalism is closely tied to populism whereas democratic capitalism rejects populism. This results in a kind of individual and national schizophrenia regarding how government should treat farmers. Articles in the democratic capitalism creed are:

1. Farming is a business.
2. Free enterprise is the most efficient allocator of resources and products; the market should decide the size and role of the farm sector in the nation's economy.
3. The family farm may be the most efficient mode of production, but if it is not the market should be allowed to function to replace it with another form of economic organization better able to supply food at low cost in national and international markets.
4. The invisible hand of the market turns private greed into public good; government should not interfere except to provide public goods and a safety net for those unable to provide for themselves.
5. Democratic capitalism is the most innovative and dynamic socioeconomic system known to humankind, promoting growth valued for itself and also for the care of disadvantaged people.
6. Workers should be rewarded according to the value of their contribution to society. This is called *commutative justice* (Brewster 1961). Providing equality of opportunity takes precedence over providing equality of outcomes.
7. The proper way to achieve status is to be proficient in one's chosen field

(Brewster 1961).

8. One who improves his or her income through honest toil is worthy of respect and emulation.
9. All persons, farmers and nonfarmers alike, are of equal intrinsic worth and dignity. Discrimination is not acceptable on the basis of occupation, race, creed, or gender.
10. The individual and family should be responsible for their economic security throughout life. To tax for purposes of redistributing wealth to nonneedy persons is institutionalized theft. An able-bodied adult who refuses to work has no inherent right to income from taxpayers.
11. No one, however wise or good, is wise or good enough to have absolute power over others. Brewster (1961) calls this the *democratic creed.*
12. Concentration of power in government, business, or labor destroys freedom of the individual. A role of government is to avoid such concentration of power, apply the rule of law, pursue sound macroeconomic policies, protect property rights and, in general, provide an atmosphere where the market will function. Property rights are inviolable and essential to preserve human freedom and to promote economic efficiency.
13. Public functions should be performed by the governmental unit closest to people and in the smallest unit of government within which costs and benefits of that function are realized.
14. Profit is the rudder that guides the economic ship of progress. Markets and profits are prized because they allocate resources efficiently to serve people in making their decisions to increase their well-being. Food for profit is food for people.
15. Everyone has a right to choose one's own occupation, to fail or succeed in that occupation without recourse to subsidies, to save and invest earnings, and to convey earnings to heirs.

It is of interest that farm fundamentalism and democratic capitalism alike claim inspiration from Thomas Jefferson. Farm fundamentalists view government commodity programs as serving Jefferson's ideal–the small farm. Democratic capitalists see today's farm dependent on government handouts as proof of Jefferson's dictum that "dependence begets subservience." They view reliance on the government in a controlled market as a mockery of Jefferson's ideal–the self-reliant, independent, and proud yeoman farmer.

**Origins of Democratic Capitalism**

Democratic capitalism has deep roots. First there is the Judeo-Christian culture of the Western world that seeks mastery over nature, multiplying the output of natural resources through technology. There is the Reformation, which emphasized success in the secular worth and dignity of all persons (priesthood of all believers), democracy in governance (authority from the bottom up), and the work ethic (excellence in one's calling). There is English democracy, with roots in Greek antiquity, the Magna Charta, and the early parliamentary system. There is capitalism with its vigor strengthened by English political stability, institutional structure (e.g., lending for interest no longer classed as usury, the limited liability corporation, etc.), and the Industrial Revolution.

Then there is *laissez faire.* It had origins in the utilitarian philosophy of Adam Smith whereby a man because of his acquisitive instinct is led to Utopia by the invisible hand of the market. There is eighteenth-century Enlightenment philosophy, which in England emphasized reason, science, empiricism, and individualism. These themes were outlined more fully in chapter 2 of this volume.

Among farm organizations, the American Farm Bureau Federation historically has most nearly embodied the principles of democratic capitalism. Its 1976 policy statement read, "Freedom of the individual versus concentration of power which would destroy freedom is the central issue of all societies" (pp. 1-4). The policy statement went on to assert that the federation believes "in the American capitalistic, private, competitive enterprise system in which property is privately owned, privately managed, and operated for a profit and individual satisfaction."

In recent years, the American Farm Bureau Federation has drifted toward populism, with strong support for major government assistance to farmers in the 2002 farm bill. Nonetheless, the federation retains the firmest support of democratic capitalism among major farm organizations (see chapter 9). It affirmed in 2002 its belief in private enterprise, property rights, and rewards "according to productive contributions to society."

Theologically conservative Protestants provide support for democratic capitalism. Merrill Oster (1987), who is a farmer, agribusinessman, and representative of conservative Christian thinking, speaks for many others when he says:

> This is the kind of society John Adams and Thomas Jefferson had in mind. The America that Adams saw is founded on certain old-fashioned concepts: that free societies produce the most good and the least evil; that theft is wrong; that work is noble and expected of everyone; that a man should be willing to stand on his own two feet; that he should be free to succeed or fail and keep the fruits of his labors; that it is not right to accept that which you have not earned; and that when we are successful we have responsibilities to voluntarily share our wealth with those who, because of youth or age or sickness or incapacity, cannot take care of themselves–especially our own families. (p. 22)
>
> We need to defend the concept of human freedom being the basis for legal and social foundations of society. This legal and social order produces the economic framework for the free enterprise system. And we must never take for granted the fact that the fabric of civilization rests on honest money, honoring of contracts, freedom coupled with responsibility, individual property rights, limited government, individual and group integrity, and living within our means personally and nationally. The result is that free choice under the free market system has unleashed technological progress that makes the American system the most developed poverty-fighting system in the world. (p. 251)

Ezra Taft Benson, secretary of agriculture under President Eisenhower, was keenly aware of the spiritual values of rural life and later became head of the Mormon Church. He noted that "the country is a good place to teach the basic virtues that have helped build the nation," but went on to place himself on the side of democratic capitalism by asserting that "agriculture is not so much an important segment of our population as of our free enterprise system. It should be permitted to operate as such" (see Carlson 1986, p. 18).

The continuity of democratic capitalism is clearly apparent from Jefferson to the modern day. According to Abraham Lincoln "my opinion of them [farmers] is that, in proportion to numbers, they are neither better nor worse than other people (Fite 1962, p. 1208). Lincoln showed no special enthusiasm, based on farm fundamentalism, for pushing through the landmark tripartite agricultural legislation of 1862: the Morrill Act establishing land-grant colleges, the Homestead Act giving access to land for thousands of families, and the act establishing the U.S. Department of Agriculture. Lincoln favored these because he subscribed to democratic capitalism–he believed that passage would benefit the nation as a whole (Fite 1962, p. 1). Lincoln's recognition of the contribution of private property to diligence under democratic capitalism is apparent in his statement (taken from Novak 1987): "When those who work in an enterprise also share in its ownership, their active commitment to the purpose of an endeavor and their participation in it are enhanced. Ownership provides incentives for diligence."

**Failings of Democratic Capitalism**

Democratic capitalism like farm fundamentalism is characterized by numerous myths. A slightly rephrased version of the 12-step drug reformer's prayer is apropos. We need the courage to use markets where they work, the serenity to accept other allocators where markets do not work, and, above all, the wisdom to know the difference. The market underinvests in public goods such as basic agricultural research and general education. The market alone will not avoid concentration of power. The market alone will not provide macroeconomic stability. Economists note, however, that the social cost of public intervention is frequently bigger than the social cost of the market failure the government was trying to correct. Thus market failure is not itself a sufficient condition to justify public intervention; populist nostrums often turn out to be costly and counterproductive.

The market alone will not provide distributive justice to all those who are physically or mentally handicapped and who in an impersonal urban-industrial society cannot rely on family or private charities. Democratic capitalism adherents emphasize equality of opportunity rather than insist on equality of outcomes. But their commitment to distributive justice, defined as providing equality of opportunity and giving able-bodied adults an equal voice in shaping rules of the economic gain, is sometimes weak and inherently in conflict with commutative justice that calls for rewards to resources according to the value of their contribution to output.

Economic outcomes under capitalism are skewed in favor of the wealthy minority. Taxing of earnings to redistribute earnings from the rich to the poor violates the concept of commutative justice. The lower wealth majority has motive and means to vote redistributions of wealth from the wealthy minority to the poorer majority. Thus, democracy and capitalism sometimes conflict. The vast majority of income transfers by government go to the middle and upper classes, partly because the middle class must be bribed to vote for transfers to the poor.

Businesses and wealthy consumers use indirect political processes such as financial contributions to parties and candidates along with lobbying to counterbalance the voting power of the populist masses. Democracy and capitalism coexist quite well despite internal contradictions. Indeed, in world perspective, democracy exists only in capitalist, free-enterprise countries. The internal contradiction must not be decisive. Nonetheless, many observers frustrated with populism would like to see the economic system be able to trump the political system rather than the other way around as with the current system.

## RESOLVING VALUE CONFLICTS

It is difficult to find a person or organization fitting precisely the farm fundamentalist model or the democratic capitalist model. However, many organizations are closer to one philosophy than the other. Conservative church groups, American Farm Bureau Federation, National Cattlemen's Association, National Association of Manufacturers, U.S. Chamber of Commerce, and the Republican Party lean to democratic capitalism. Liberal church groups (including some, such as the Catholic Church, that are liberal on economic issues but conservative in social issues and theology), the National Farmers Union, American Agriculture Movement, National Farmers Organization, various crop-growers associations, AFL-CIO, and the Democratic Party tend toward farm fundamentalism.

Farm fundamentalists do not reject the market, but believe that relieving farm hardship and preserving the family farm is more important than maintaining an open market. Democratic capitalists on the other hand believe that farmers and the rest of society will be better off economically in the long run by facing the impersonal rigors of the market. If family farmers are unable to compete, then society is best served in the long term by letting inefficient or unlucky farmers go the way of the blacksmith, buggy maker, and the mom-and-pop grocery store.

There is much psychological value in having a creed to live by whether that creed be myth or reality. Farm fundamentalism, for example, meets a deep-seated need in farm people for a feeling of worth and dignity. In nonfarm people, farm fundamentalism responds to the longing for a nostalgic, romantic, unifying symbol of a simpler and less hectic bucolic life and heritage.

Myths can be dangerous, however. Farm fundamentalism, unfortunately, has brought more concern for the family farm than for the farm family. Farm fundamentalists have been so preoccupied with preserving the family farm that they have not had time or inclination to prepare farm people released from farming by laborsaving technology for their inevitable adjustment to nonfarm employment either as part-time farmers or as town or city residents. Billions of taxpayers' dollars spent each year have not preserved the family farm. Funds would have had much higher payoff if spent for vocational-technical and general schooling, job information, mobility assistance, and personal and financial counseling.

Many youth have received useful training in responsibility, leadership, and citizenship in the Future Farmers of America. The downside is that such training can distract students from obtaining strong grounding in mathematics, literacy, and science basic for achievement in an urban postindustrial society. Vocational training in farming imparted unrealizable expectations of farming as an occupation for many youth. In an environment of farm fundamentalism, students were not exposed to the reality of huge capital, management, and risk-taking requirements essential for a successful family farm operator. Once established in farming, poorly prepared young operators were unable to continue in the face of even modest economic setbacks.

## OTHER COSTS OF POPULISM OPERATING THROUGH FARM FUNDAMENTALISM

A theme of this volume is that agricultural radicalism and populism are costly. This final section of chapter 7 builds the case that populism is as costly as radicalism. From 1933 to 2000, taxpayers spent $561 billion (year 2000 dollars) just on farm price and income supports (Tweeten 2002). Federal outlays for farm programs averaging over $20 billion per year in the late-1990s and early-2000s were mostly transfer payments from taxpayers to commercial farm operators and ultimately to landowners. But "real costs" in lost national income were also sizable.

Market distortions caused by commodity programs reduced national income in several ways. Federal income taxes used to support government programs lose on average at least $16 of real national income (deadweight loss) per $100 collected (Ballard, Shoven, and Whalley 1985, p. 13). Thus, government payments to farmers averaging $18.6 billion per year from 1998 to 2000 lost $3.0 billion of national income annually because the public made different savings, investment, and labor-use decisions than it would have in the absence of taxes that finance farm programs. The 2002 farm bill commits to continue spending nearly $20 billion annually on farm programs for another decade.

Excessive output and resources committed to farming cost the nation an additional $0.93 billion in lost income based on the average, 3.4 percent, estimate of uneconomic production in excess of actual output of grains, soybeans, and cotton from 1998 to 2000 (deadweight costs estimated by Tweeten [2002]). Adding administrative and lobbying resources (about $2 billion that could have been better used elsewhere) brings the annual total loss in national income to $5.93 billion. Further accounting for national income loss of $600 million from peanut, tobacco, sugar, and dairy programs brings the total loss of national income due to farm programs to $6.53 billion, or 3 percent of farm receipts from 1998 to 2000. This estimate from Tweeten (2002) understates the full cost because is does not count the cost to farmers of their considerable time and effort devoted to exploiting the system to their advantage. Also, the estimate is static and does not consider the longer-term general equilibrium losses in national income from less savings, investment, and productivity. General equilibrium models indicate that accounting for dynamic impacts could double the static estimate, $6.53 billion, of national income forgone each year due to farm commodity programs.

At least two issues must be confronted before attributing the costs to farm fundamentalism and its populist manifestation in politics. One issue is whether benefits of commodity programs offset the costs. The second issue is whether farm fundamentalism played a role in the politics of support for farm programs.

Using traditional welfare economics criteria that government interventions serve people by serving economic equity, efficiency, stability, and freedom goals, I am unable to build any economic case for commodity programs other than as a transition to ease farmers away from dependency (Tweeten 2002). It is not that programs have not been useful such as in saving family farms in the early 1980s. Rather, it is that other means could have been used to better accomplish the same goals at a fraction of the cost. Transfers of tax dollars to the commercial farmers and farmland owners with wealth and income well above that of the average American household is questionable policy at best if the goal of public policy is to promote well-being of people.

Turning now to the role of farm fundamentalism in the making of farm policy, Orden, Paarlberg, and Roe state that "we found that agrarian mythology [fundamentalism?] played little or no role in the 1996 FAIR Act outcome" (1999, p. 227). That may be what they observed but the power of agrarian fundamentalism is not directly observed. It is the "800-pound gorilla" that need not speak to have influence. It explains why the 99.8 percent of the nation's population that pays the tab for commodity programs lets the members of Congress who represent commercial farm interests have their way. The 150,000 largest farms account for two-thirds of transfer payments received from taxpayers, but those farms account for only 7 percent of farmers and for only 0.2 percent of the nation's population. This tiny segment of the population does not have voting muscle to dictate farm policy.

Other political factors also influenced Congress to fund farm commodity programs at an unprecedented pace in the late-1990s and beyond when per household income and wealth of farmers well exceeded those of nonfarm households. Republicans and Democrats were in hotly contested races for presidential and congressional control. They engaged in a bidding war for farm programs and votes. That's an attractive strategy indeed for politicians when the race is tight and other people's (taxpayers') money is being spent.

Despite the high cost, farm votes tend to be cheaper to get than other votes. Farmers are thought to be switch voters, willing to vote for politicians who respond to their "pocketbook" needs. Many states with small populations are farm states with more seats in Congress and electoral votes per capita than large urban states. The strategy worked for normally conservative Republicans bidding against populist Democrats, giving Republicans the Congress and presidency in the 2000 election. The bidding war could continue because it is a "prisoners' dilemma" –the nation would be better off without it but each party cannot afford to stop bidding.

## REFERENCES

American Farm Bureau Federation. *Farm Bureau Policies for 1976.* Resolutions adopted at 57th annual meeting in St. Louis, MO. Chicago: AFBF, 1976.

Ballard, C. L., J. B. Shoven, and J. Whalley. "General Equilibrium and Computations of the Marginal Welfare Costs of Taxes in the United States." *American Economic Review* 75 (1985): 128-38.

Brewster, J. "Society Values and Goals in Agriculture." In *Goals and Values in Agricultural Policy*, Center for Agricultural and Economic Development. Ames: Iowa State University Press, 1961.

Burchinal, L. "The Rural Family of the Future." In *Our Changing Rural Society,* edited by J. Copp. Ames: Iowa State University Press, 1964.

Carlson, A. "Should America Save Its Peasant Class?" *Small Farmer's Journal* 10, (Winter 1986).

Coughenour, C. M., and L. Tweeten. "Quality of Life Perceptions and Farm Structure." In *Agricultural Change,* edited by Joseph Molnar. Boulder, CO: Westview Press, 1986.

Dingman, M. "What Does Christianity Have to Do with the Farm Crisis?" Proceedings of the conference *Is There a Moral Obligation to Save the Family Farm?* Ames: Religious Studies Program, Iowa State University, 1986.

Drury, R., and L. Tweeten. "Have Farmers Lost Their Uniqueness?" *Review of Agricultural Economics* 19 (Spring/Summer 1997): 58-90.

Fite, G. "The Historical Development of Agricultural Fundamentalism in the 19th Century." *Journal of Farm Economics* 44 (1962): 1203-11.

Goldschmidt, W. *As You Sow*. Mountclair, NJ: Allenheld, Osmun, and Company, 1946; reprint 1978.

Hamilton, H. "Social Implications of the Family Farmer." *Family Farm Policy: Proceedings of a Conference on Family Farm Policy*, edited by J. Ackerman and M. Harris. Chicago: University of Chicago Press, 1946.

Heady, E. "Preface." In *Goals and Values in Agricultural Policy,* Center for Agricultural and Economic Development. Ames: Iowa State University Press, 1961.

Jefferson, T. "Letters to Jay," August 23, 1785. In *The Best Letters of Thomas Jefferson,* edited by J. G. Hamilton. Boston and New York: Houghton Mifflin Co., 1926.

Jefferson, T. *Notes on the State of Virginia.* Philadelphia: Prichard and Hall, 1788.

Jordan, B., and L. Tweeten. *Public Perceptions of Farm Problems*. Research Report P-894. Stillwater: Agricultural Experiment Station, Oklahoma State University, 1987.

Orden, D., R. Paarlberg, and T. Roe. *Policy Reform in American Agriculture*. Chicago: University of Chicago Press, 1999.

National Issues Forum. "America Looks at Agriculture: An Analysis of Contemporary Attitudes on Some Basic Issues." Briefing Book No.1. Helena, MT: AgFocus, Office of the Governor, 1986.

National Issues Forum. "The Farm Crisis: Who's in Trouble, How to Respond." Briefing Book No.2. Helena, MT: AgFocus, Office of the Governor, 1987.

Novak, M. "Cash Income and the Family Farm." In *Are Farmers Exploited by Agribusiness Corporations?,* edited by G. Comstock. Ames: Religious Studies Program, Iowa State University, February 1987.

Oster, M. "Do Religious Values Suggest 'Family Farms' are More Socially Desirable than 'Corporate Farms'?" In *Is There a Moral Obligation to Save the Family Farm?,* edited by G. Comstock. Ames: Iowa State University Press, 1987.

Paarlberg, D. *American Farm Policy.* New York: Wiley, 1964.

Pasley, J. "The Idiocy of Rural Life." *New Republic*, December 8, 1986 , pp. 24-27.

Persaud, S., and L. Tweeten. "Impact of Agribusiness Market Power on Farmers." In *Agricultural Policy for the 21st Century,* edited by L. Tweeten and S. Thompson. Ames: Iowa State Press, 2002.

Pius XII. "Problems of Rural Life." *Christianity and the Land.* Des Moines: National Catholic Rural Life Conference, 1951.

Pius XII. "La Solennita della Pentecoste," June 1, 1941. In *Principles for Peace,* edited by H. C. Hoenig. Washington, DC: National Catholic Welfare Conference, 1943.

Speltz, G. "Theology of Rural Life: A Catholic Perspective." In *Farm Goals in Conflict,* Center for Agricultural and Economic Development,. Ames: Iowa State University Press, 1963.

Tweeten, L. *Farm Policy Analysis*. Boulder, CO: Westview Press, 1989.

Tweeten, L. "Farm Commodity Programs: Essential Safety Net or Corporate Welfare?" In *Agricultural Policy for the 21st Century,* edited by L. Tweeten and S. Thompson . Ames: Iowa State Press, 2002.

Zimmerman, C. "The Family Farm." *Rural Sociology* 15 (September 1950): 211-21.

# 8

# Populist Mythology

Agricultural populists present an image of agriculture supporting their argument for transfers of income from consumers and taxpayers. In contrast to radical agriculturalists who contend that problems in food and agriculture arise from imbalances between humans and nature, populists contend that farm problems arise from economic imbalances. The most glaring imbalance is the "unjustly low" income share going to farmers versus other sectors of the economy.

Even as Americans labor mightily to convince developing and former communist countries that markets work, populists continue to convince politicians that markets don't work in U.S. agriculture. Farm populists have felt that belief so strongly that from time to time they have protested violently against the "system." Chapter 9 provides historic perspective on farm organizations that emerged from protest movements.

The purpose of this chapter is to review the validity of some of the populist premises on which agricultural policy is based. The ten myths perpetrated by agricultural populists scrutinized in this chapter build on issues addressed in Chapter 7. The set by no means exhausts the mythology of populism (Box 8.1).

## SELECTED MYTHS

Populists, if anything. are resourceful and innovative, concocting myths as need arises. Myths relate to technology, markets, prices, and the nature of competition.

### Myth 1: All wealth is from raw materials.

Economic theory does not intellectually justify the demands of agricultural populists. So populists have had to concoct their own conceptual framework. One of the most basic and amusing populist myths is the self-serving *raw material theory of value.* It justifies special treatment for agriculture to raise farm income *and national income.*

**Box 8.1**

The market does its magic creating wealth by rewarding innovative, efficient firms and by culling laggards. Joseph Schumpeter called this process of continuing renewal "creative destruction." The marginal commercial farmers prominent in populist movements find such a message anathema because they fear they will be culled. Barry Flinchbaugh, an able extension economist at Kansas State University, has frequently been candid in saying how a successful economy operates–most recently in presenting ten farm myths, which for the most part do not overlap with those in this chapter (Ehmke 2002, p. 16). Over the years, his populist critics have responded. After receiving threatening phone calls, he has had on occasion to be escorted by law officers around the state. He was hanged in effigy on the Kansas State University campus in 1986 by a mob of farmers and students. A wheelbarrow of horse manure was heaped in front of his office after he stated correctly that some farmers were making money.

Farm populists have found a role model in organized labor–in its labor-supply control, high negotiated wages, and labor theory of value. For farm populists (mostly farmers on midsized, commercial family farms), the farm equivalents are commodity supply management, "parity" commodity prices, and the raw material theory of value. The policies, if implemented, that would follow from these myths have enormous capacity for economic mischief.

The central principle of the *raw material theory of value* is that raw materials are the sole source of the nation's wealth. "Raw materials" are products of primary industries such as agriculture whose output is extracted from natural resources. Corollaries are that (1) pricing of raw materials at "full" or "honest" parity (never clearly defined but interpreted by many to mean the same ratio of prices received to prices paid by farmers as prevailed in the 1910-14 period) will restore and sustain the nation's economic health, and (2) national income is a simple multiple of income from raw materials.

The doctrine traces to the eighteenth-century Physiocrats of France who claimed that national wealth was a multiple of agricultural wealth. The mathematical relationship was expressed in the *Tableau Economique* of François Quesnay. The doctrine, long interred by neoclassical economics, was exhumed and powerfully advocated by Carl H. Wilken in the 1930s. The theory was strongly advocated by Charles Walters, Jr., in his book *Unforgiven* (1971). It is the *raison d'être* and current core teaching of the National Organization for Raw Materials (NORM 2002) founded over 40 years ago by Carl Wilken.

A key issue is whether in fact raw materials determine the value of a commodity. The reverse of the proposition is correct: the value of a raw material (or other resource) is determined by its contribution to the value of the commodity it produces. That value has two components: the contribution of the resource to output and the value of that output (commodity) to consumers. The value to consumers depends much more on nonmaterial attributes such as love, peace, beauty, information, knowledge, and taste than on material attributes of the good or service. The market clears at a commodity price at which producers are willing to supply the market with exactly as much of the commodity as consumers demand at that price.

These "laws" of supply and demand apply in barter, free enterprise, and socialistic economies–the laws can no more be repealed than the law of gravity. To be sure, pricing can deviate from the competitive market level when it is administered or negotiated by individuals, firms, and agencies that can control industry supply and/or demand. Such manipulation is widespread but at a real cost to society in lost value of goods and services produced and consumed as we observed for commodity programs in the last chapter.

The real value of goods and services in an economy is greatest with competitive pricing and output; hence the competitive norm is useful for measuring performance of the economy, *especially* one characterized by much imperfect competition. Competition need not be atomistic (many firms) or pure (perfect mobility, knowledge, etc.) for efficient market performance. In the case of Boeing and Airbus or Pepsi and Coke, two firms provide workable competition for high performance. The scope for exploitation by imperfectly competitive firms is sharply curtailed by free international trade that subjects firms to competition from abroad.

NORM and other populist organizations maintain that 100 percent of farm price parity will make not only agriculture but also the entire economy bloom. The leverage principle of $1 of farm income creating $7 of national income grew out of the 1930s when farm receipts happened to be one-seventh of national income. That multiplier supports making agriculture prosperous by whatever means necessary to make the nation prosperous.

That multiplier went from $7 in the 1930s to $70 in 2001, a gain greeted with approbation by true believers but with ridicule by most everyone else. In 2001 national income was 70 times gross farm income. Are we to believe that raising farm income by $1 will raise national income $70? The rising multiplier is easily explained. As consumers' income rise, they choose to spend a declining proportion of it for food and fiber ingredients supplied by farmers. This pattern of declining demand for farm products relative to nonfarm products with economic growth is not unique to this country–it is a worldwide phenomenon. The pattern may appear insidious to farmers but is a normal expression of consumers' preferences. It follows that national economic growth reduces the ratio of farm sector income to national income.

Economists use the concept of basic and nonbasic industry, where basic industry is defined as that which brings dollars to a region or area from outside. As such, farming is a basic industry that generates nonbasic input-supply, marketing, and consumer service industries. Each dollar of additional farm income in a state tends to generate approximately $2 of income in nonbasic industries. Thus, the total income multiplier for farming is approximately $3. This multiplier is an "engineering" coefficient useful for determining the impact of a change in farm income on area or regional income. The multiplier does not reveal whether additional farm output is economically feasible or worthwhile! If farm output is in excess supply, consumers are satiated, and product price is below the cost of production, additional output brings economic loss to farmers and society. Forcing the additional output to be delivered to consumers wastes farm and marketing resources valued more highly by society in other uses.

In its simplicity the raw material theory of value is akin to the labor theory of value originated by David Ricardo and later embraced by Karl Marx. The labor theory of value attributing all wealth creation to labor is absurd and is unable to explain how goods and services are valued. Few goods and services arise from input of labor alone or of raw materials alone. The fourfold increase in farm output in the twentieth century originated from nonconventional inputs–science and technology–rather than from conventional labor or raw material natural resources (see Council of Economic Advisors 2001, p. 389, and earlier issues). The nonconventional inputs in turn were largely the result of application of human capital in turn generated by education.

**Myth 2: Farm commodity markets don't work.**

Myth 2 holds that, absent interventions in markets by government or farmers wielding market power, the farm economy will be troubled by chronic excess production and by low commodity prices, incomes, and rates of return on investment. Farmers, it is said, do not adjust resources fast enough in response to rapidly changing technology to avoid perennial low rewards for their resources. That is, new, technologically improved capital resources enter farming to increase output and reduce farm prices and receipts (due to an inelastic demand for food). Hence, less-productive farm resources, mostly labor, must move to nonfarm employment if returns on farm resources are to remain favorable.

Farming productivity has advanced at about the same rate as demand for farm output for decades so that aggregate farm resources have not had to leave farming. The nation has nearly the same real volume of resources in farming today as it had in 1910-14 (Council of Economic Advisors 2001, p. 389; U.S. Department of Agriculture 1980, Table 69)! The mix of resources has shifted from reliance on labor to reliance on capital, however.

So we must look to whether farmers have been able to change the mix of labor and capital as fast as required to maintain favorable returns on farming resources. The short answer is that, after lagging adjustments since the 1930s, farm labor outmovement has been able to keep up with adjustment pressures on average. Since about 1970, agricultural resource returns have been at levels consistent with a well-functioning market except for some cyclical and annual deviations. The latter phenomenon is a problem of instability and not a problem of chronic low returns on resources.

A well-functioning market maximizes real national income, and rewards competent farm operators of adequate size farms with a return comparable to what those resources could earn elsewhere. If "adequate size" commercial farms are those with sales of $250,000 or more in 1997, then rates of return on assets averaged over 7 percent (fiftieth percentile as measured by Hopkins and Morehart [2002]). If "able" operators are the top half of commercial farmers, then the rate of return on farm assets averaged near 20 percent in 1997 (seventy-fifth percentile as measured by Hopkins and Morehart [2002]). Of course, the market will not reward incompetent operators of undersized farms with favorable rates of return any more than it will award full-timer earnings to part-timers or incompetents in other occupations.

Farm household income averaged well above that of nonfarm households from 1996 to 2001, exceeding nonfarm household income by 16 percent per household in 1998 and 1999 (U.S. Department of Agriculture October 2001, p. 49). Farm household income from nonfarm sources alone has exceeded income per nonfarm household in recent years! Farm net worth per household averages about double the net worth per nonfarm household.

Farmers' economic success traces to technology making possible the growth of farm size and off-farm income (Tweeten 1994, p. 7). Economic success does not trace to government commodity programs: rates of return, income, and net worth do not seem to differ much between farms that participate or do not participate in commodity programs (see Tweeten 1989, ch. 4).

**Myth 3: Farmers need 100 percent of parity to earn a fair return.**
The National Farmers Union, National Farmers Organization, American Agriculture Movement, and other populist farm groups have labored long to bring government intervention in markets to achieve 100 percent price parity by supply controls if necessary. However, aggregate farm commodity price or terms of trade (*parity ratio)* defined as the ratio of the index of prices received by farmers for crops and livestock to the index of prices paid by farmers for inputs–with the ratio expressed as a percent of that in the 1910-14 period–is a poor measure of market efficiency and fairness.

Farm commodity terms of trade or parity in 1999 were only 40 percent of the 1910-14 average (U.S. Department of Agriculture October 2001, p. 29). This number has been interpreted by some to mean that farmers are only receiving 40 percent of a "fair" price. In fact, as explained below, real farm commodity prices have risen!

The multifactor productivity index (ratio of aggregate output of crops and livestock to aggregate farm production input) in 1999 was 3.94 times the 1910-14 level. In other words, through the help of science and agribusiness, farmers were able to grow four "blades of grass" where only one grew in 1910-14! Thus, the factor terms of trade index, defined as the real price (purchasing power) of farm output per unit of farm production input, was 158 percent (40 x 3.94) of the 1910-14 level. It follows that the real buying power (factor terms of trade) of the average production input was 58 percent greater in 1999 than in the 1910-14 base period!

Alternatively, farmers "growing 4 blades of grass where one grew" in 1910-14 with the same input volume means that farmers needed only 25 percent as high a real price per "blade" in 1999 to achieve the same real income per unit of production input as in 1910-14. In fact, however, because farm commodity prices in 1999 averaged 40 percent of those in 1910-14, real prices received per factor input in 1999 averaged 40/25 or 158 percent of prices in 1910-14 after the parity ratio is adjusted for productivity growth. It follows that, comparing 1999 with 1910-14, the real parity ratio (factor terms of trade) was *up* by 58 percent rather than down 60 percent as implied by the conventional parity ratio (commodity terms of trade) of 40 percent.

Benefits to the economy have been widely shared among those who left the farm as well as those who stayed–both groups raised their incomes and living standards. Some persons were "left behind" in the farm *and* nonfarm sectors, but these persons will be helped by public assistance, job training and information, and relocation assistance rather than by 100 percent of farm price parity.

Another favorite populist nostrum, cost of production pricing, has as many pitfalls as parity pricing. In a major shift from its historic support for supply controls to hold farm prices at some (high) percent of *parity*, the populist National Farmers Union in 2002 farm bill deliberations called for prices supported to cover the *cost of production*. Production controls would set output to bring one price for all farms. Because price could not be set to precisely cover cost on each farm, whose production cost is to be used to set price? Small farms whose production costs are high? Price supported at that high level would provide huge windfall gains to larger, more efficient farms, would lift land prices to form a formidable entry barrier to new operators of all farms, would raise production costs to lift price supports ever higher, and would provide windfall gains enabling large farms to buy up small farms.

On the other hand, prices supported to cover the lower costs of large farms would not help small farms but still would induce additional output. A reason is because the average (expected) price given price support at P exceeds the expected market price known to average P. In other words, cutting off the lower tail of the price distribution raises the *expected* average commodity price, thereby raising production and lowering market prices.

Price set by government at 100 percent of parity or some other level above that of a competitive market does not raise farm income because the support stimulates supply to reduce market price, retards consumption as consumers switch to less-expensive goods, and is lost through capitalization into rents and land values. Benefits bid into rents and land prices are gained by landowners and are lost to renters and new landowners.

Land prices (inflated by past programs) could fall 25 percent with termination of farm commodity programs (Tweeten 2002). Thus, past programs have created a moral hazard requiring a carefully phased withdrawal of programs to avoid a farm financial debacle. Incentives to overproduction under the 1996 farm bill reduced farm receipts to offset payments (Tweeten 2002); the 2002 farm bill promises more of the same.

**Myth 4: Technological change has made farmers worse off.**

Technology is not the villain it is often portrayed to be. It has raised real economic terms of trade for farmers and nonfarmers alike–including for farmers who left their operations in midcareer. Technology not only helped real farm prices to rise as noted above, it also helped real farm income to improve. Real personal income per capita of the U.S. population increased to 3.5 times its 1930 level by 1999, an average annual gain of 1.8 percent per year (Council of Economic Advisors 2000,

p. 335, and earlier issues). Meanwhile, per capita income of farmers increased from 40 percent that of the average American to 117 percent of the average American (U.S. Department of Agriculture 1960, p. 38; March 2001, p. 46). It follows that real income per person on farms was approximately 10 times (3.5 x 117/40) higher in 1999 than in 1930. Government payments had little impact on that income gain.

Agriculture has contributed mightily to the rise in the nation's real income and standard of living. Because income of farm households tends to be equal to (or a little better than, as shown earlier) income of nonfarm households in our well-functioning economy, it follows that measures that increase productivity in agriculture or elsewhere will eventually be spread throughout the economy to benefit most people. That outcome is explained by the fact that markets robustly move household income to equality among sectors as demonstrated by history.

Moving to that equilibrium has meant moving some farm people to nonfarm occupations more favored by demand as national income rises. Leaving a farm in midcareer can be more traumatic than leaving other employment because the farm is a way of life and a home as well as a job. Nonetheless, farm operators and their families seem to adjust well for the most part, perhaps because a high proportion of farm people have experienced off-farm employment before they leave the farm. A 1987 survey in Oklahoma of 295 midcareer farm leavers found that those who said they are somewhat or much better off outnumbered those who said they are somewhat or much worse off by a three to one margin (Perry, Schreiner, and Tweeten 1991). Two-thirds of the 624 former farmers in North Dakota interviewed in 1986 said they were better off since they quit farming (Bentley et al. 1989).

**Myth 5: Off-farm income should not be included when measuring the economic welfare of farm households.**

Only $5,354 of farm household income averaging $61,796 annually from 1998 to 2000 was from farming activities (U.S. Department of Agriculture October 2001, p. 49). Populists have noted the small average income from farming per household, and have used the statistic to justify continuation of federal farm income support programs. That justification is seriously flawed for several reasons.

- As Professor Barry Flinchbaugh of Kansas State University has frequently asserted, "Bankers can't tell the difference between a dollar coming from the farm and a dollar coming from elsewhere."
- Workers in nonfarm American families regularly engage in multiple jobholding. Farmers do the same. Off-farm work serves commercial farmers well by relieving intense cash-flow problems of debt service and living expenses for family farms refinanced every generation. Off-farm work is the core of economic existence for hobby farmers enabled to enjoy the amenities of a rural landscape and way of life.
- Finally, *average* income from the farm of farm households is a nearly meaningless number. It lumps together the average (1) farm losses of the three-fifths of farmers, mostly small, part-time hobby farmers (who reside

**Table 8.1. U.S. Farm Household Numbers, Government Payments, and Income by Farm Category, 1998**

| | Farm household income | | | | Share of | | |
|---|---|---|---|---|---|---|---|
| Farm category | Government payments | Crops & livestock | Total from farm | Total all sources | Operator's households | Government payments | Farm receipts |
| | Dollars per household | | | | Percent | | |
| All operator households | 4,291 | 2,234 | 6,525 | 59,734 | 100.0 | 100.0 | 100.0 |
| *Small Family Farms* | | | | | | | |
| Limited-resource | 620 | -3,849 | -3,229 | 9,924 | 5.8 | 0.9 | 0.6 |
| Retirement (retired) | 1,494 | -2,993 | -1,499 | 45,659 | 13.6 | 3.3 | 1.4 |
| Residential/lifestyle (nonfarm occupation) | 902 | -5,211 | -4,309 | 72,081 | 42.6 | 8.6 | 6.1 |
| Farming occupation: | | | | | | | |
| Lower sales (under $100,000) | 2,260 | -4,673 | -2,413 | 34,773 | 22.0 | 14.5 | 7.8 |
| Higher-sales ($100,000-249,000) | 11,314 | 10,149 | 21,463 | 50,180 | 8.0 | 25.3 | 17.1 |
| *Large Family Farms* ($250,000-499,000) | 21,451 | 37,838 | 59,289 | 106,451 | 3.5 | 20.8 | 16.8 |
| *Very Large Family Farms* (over $500,000) | 25,379 | 150,486 | 175,865 | 209,105 | 2.7 | 22.5 | 36.7 |
| *Institutional Farms* | NA | NA | NA | NA | 1.8 | 4.1 | 13.6 |

*Source*: U.S. Department of Agriculture, September 1999, February 2000.

on the farm to realize rural amenity and tax write-off benefits), and (2) sizable positive income from farm sources earned by commercial operators who supply most farm output.

A related issue is the National Farmers Union position that farm income is better measured by the Internal Revenue Service (IRS) than by the Economic Research Service (ERS) of the U.S. Department of Agriculture. I used data provided by the U.S. Department of Agriculture and other sources to reconcile the difference between the IRS net farm income estimate of $4.20 billion and the ERS estimate of $46.90 billion in 1989, a year when the Farmers Union was especially pressing the issue (Tweeten 1994, p. 30). The IRS shortfall was accounted for largely by unreported farm income ($11.30 billion), exclusion of farms (institutional farms such as research station, etc.) not filing IRS 1040F forms, and by farm income tax advantages such as farm dwelling services, livestock reported as capital gains rather than ordinary income, products produced and consumed on the farm, and the like. The Farmers Union used IRS data to make a case for farm commodity programs to help an "economically destitute" sector. Farmers' success in securing from Congress favorable tax treatment making their income look low may be the envy of other economic sectors, but for farm populists to then attempt to "double dip" by securing subsidies to compensate for the contrived low income is chutzpah.

**Myth 6: High risk unique to farming justifies commodity programs.**
Farming is risky, but risk is not in itself a case for subsidies from taxpayers. We don't provide government payments to lottery players, Las Vegas gamblers, Wall Street plungers, futures market speculators, day traders, and small businesses. Each of these faces more risk than do farmers (see Goodwin [2000] for small business risk). Participants in each of these activities voluntarily elected to face that risk.

Still, many consumers are nervous about relying strictly on the private sector to supply risk management tools to farmers and to assure plentiful food reserves for consumers. The nation produces about 20 percent more food than it consumes. The surplus that is exported is a ready reserve, insuring that American consumers will not run short of food even under the most adverse weather or wartime emergency. Food that is fed to livestock as well as the livestock themselves also could be consumed in an emergency, providing even more food security.

But shouldn't the public subsidize holding of grain buffer stock reserves to reassure nervous consumers and to stabilize farm prices? The answer is probably no. The social cost of a private sector supplying less than "optimal" amounts of reserve stocks, insurance, and forward pricing is probably less than the administrative and mismanagement costs of public stabilization policies.

Langemeier and Patrick (1990) show that farmers are remarkably good at stabilizing consumption despite unstable income from year to year. Farmers are adept at self-insurance, are not very risk averse on average, and nowhere in the world are willing to pay for unsubsidized all-risk crop insurance (Wright 1995, p. 30). Operators voluntarily enter farming and assume risks they well know characterize the industry.

Current commodity programs would have to be restructured massively to cost-effectively address problems of instability. Such restructuring would recognize that most small farms have adjusted to risk relying on off-farm income to stabilize their finances, and commercial farms on average have sufficient wealth to pay for the many private risk-management tools available to them.

Farmers have at their disposal a host of impressive risk-management tools from the private sector. These include insurance, forward markets, and storage. If these tools are deemed inadequate, then midsized family farms that frequently are least able to cope with risk can be provided with a publicly assisted risk safety net most cost-effectively by focusing stability on the "bottom line," net income, rather than on price, yield, gross revenue, or cost components of income that can vary to offset each other. An investment retirement account (IRA) type program with the government partially matching a farmer's contribution and giving tax-exempt status to interest revenue is an option to address farming instability at minimal cost, including the low transaction cost of administration by the Internal Revenue Service. That type of program could easily be extended to all farmers and regions and need not be restricted to crops. The program, recognizing that farming is not a low income/wealth sector, can focus on facilitating the shift of earnings of farm people from high- to low-income years.

**Myth 7: Farm commodity programs save small family farms.**
Table 8.1 provides helpful insights into how small farms share in commodity programs. In 1998, the 150,268 limited-resource households had income averaging $9,924, net worth $78,718, and losses averaging $3,849 from crop and livestock income. These disadvantaged households accounted for 5.8 percent of farm households and 0.6 percent of farm receipts but for only 0.9 percent of farm program payments in 1998 (table 8.1). Payments averaging only $620 per household were not only small but were only 14 percent of the average payment, $4,291, for the 2 million operator households in 1998. In contrast, very large farms (sales over $500,000 annually) received government payments averaging $25,379 and total income of $209,105 and net worth of $1,508,151 per household. Although large and very large farms have especially large payments per farm, it is interesting to note in table 8.1 that very large farms averaged only a little larger payments in 1998 than large farms.

Farms can usefully be divided into two broad groups in table 8.1. The first is farms whose share of government payments fell short of their share of farm households; hence they received lower payments per household than did other farm households. The second is farms whose share of government payments fell short of their share of farm receipts; hence they received lower payments per dollar of farm receipts than did other farms. Small farms, the first four categories of farms in table 8.1, all fall in the first category. None of the small farm categories fell into the second group.

Small payments per household do not necessarily contribute to economies of size and hence to fewer and larger farms if payments are large per dollar of farm receipts. The higher payments per dollar of farm receipts on small and intermediate size farms imply that farm payments reduced economies of size by leveling the cost of production per dollar of receipts (including payments) among farms. In table 8.1, the very high ratio of payments share to receipts share on "higher-sales small farms" and on "large family farms" indicates that commodity programs have helped intermediate-sized farms relatively more than smaller or larger farms, although payment per farm is greater on larger units as noted in table 8.1.

Payment limitations are not binding and program participation is high among all categories of farms, hence it may seem surprising that larger farms have such low payments per dollar of sales. The reason is that larger farms produce a disproportionate share of commodities not covered by programs–livestock, fruits, and vegetables.

The conclusion from data in table 8.1 is that government commodity programs relatively advantage intermediate-sized farms because they receive especially high payments per dollar of commodity receipts. Large and very large farms are advantaged by especially large payments per farm. Furthermore, the capital and security offered by programs allows larger farms to leverage equity to buy out and consolidate their small-farm neighbor.

Government program payments are very small for limited resource farms–the category that is most nearly a welfare case. An end to or a major restructuring of commodity programs would be necessary to help the smallest farms. Small farms dominate farm numbers. Because populist agriculture dominates farm politics and represents midsized rather than small or very large farms, commodity programs are targeted pretty well to suit them. The vulnerable midsized farms most dependent on programs are not numerous, however, and they could be assisted at relatively low cost by targeting credit programs rather than commodity programs to them.

**Myth 8: The nation will run out of farm operators.**

Populist rhetoric notwithstanding, the nation is as unlikely to run out of farm operators as it is to run out of food. Farm numbers dropped only 0.1 percent annually from 1992 to 1997, the most recent years for which census data are available. Approximately 2 percent of farm operators retire or die each year. It follows that many operators were entering farming to replace exiting operators. Of course, 1992-97 may have been a favorable time for farming starts and may not be repeated.

The entry of many operators into farming seems incongruent given that a typical family farm economic unit requires control of some $2 million of assets. The family farm would likely disappear in a generation without generous parents who help young entrants, but generous parents are here to stay. Only 5,337 operators are expected to retire or die each year on average in the next decade (Tweeten and Zulauf 1995, p. 23). Although the number of farm youth who are potential replacements is down from previous years and hence opportunities to farm are greater for them, potential sources are plentiful of commercial operators who will produce most farm output. They can come from smaller farms and from off-farm sources (Tweeten and Zulauf 1995). Although studies (Sanford et al. 1984) show relatively little movement of operators between large and small farms, not much movement is needed.

The nation is in no jeopardy from running out of food or farmers with or without commodity programs although some marginal farmers would exit farming in the absence of government subsidies. In 1999, some 85,000 farms (4.2 percent) were classified as vulnerable to financial failure, having negative farm cash flow and a debt-asset ratio over 40 percent (U.S. Department of Agriculture September 2000, pp. 20, 21). However, off-farm income would sustain many if not most of these farms in the absence of a public safety net. Small farms will survive on off-farm income and most large farms will survive on their substantial income, wealth, and production efficiency. The staying power of these farms is indicated by the fact that small-farm (10 to 180 acres) and large-farm (over 500 acres) numbers increased from 1992 to 1997 (NASS 1999, p. 10). Many financially fragile farms will fail eventually with or without farm safety-net programs.

**Myth 9: Production contracting is destroying the family farm.**

In the mind of populists, production contracting is the most onerous form of vertical coordination. Under such contracting, the contractor, usually a feed dealer or food processor, provides the chicks or piglets, the feed, the organizational management, veterinarian services, and markets. The farmer (called a grower or producer) provides the building and equipment along with the labor and day-to-day management. The advantage to the contractor is to get the right product at the right time at the right price at the right quality to meet the demands of the consumer and to keep the feed mill and processing plant busy. The advantage to the farmer is access to credit (bankers will lend for buildings and equipment if the grower has a multiyear contract and guaranteed placements) and to technology and know-how while reducing capital requirements and risk.

Production contracting first became prominent in the 1950s. Hundreds of thousands of farmers in the South were on units unable to compete with other parts of the nation because their rough terrain was unsuited for mechanization that was sweeping the nation. Few off-farm jobs were available without moving to urban areas. Being poor and short on capital, their economic options were few. Availability of contracts to produce broilers allowed them to supplement their incomes and preserve their small farms and lifestyles.

Similar success was apparent elsewhere. In Mercer County, Ohio, for example, production contracting is ubiquitous in hogs, broilers, turkeys, and laying hens. Mercer County farm operators averaged farm receipts of $202,740 versus $108,193 per farm in nearby Van Wert County in 1998 (Tweeten and Flora 2001, p. 26). Soils and topography are similar between the counties, but cash-crop Van Wert County supported only 2.1 farms per square mile whereas production-contracting Mercer County supported 3.4 farms per square mile. The father of one of my students from Mercer County (who farmed with his son) had a 72-head dairy cow independent operation and a 150,000 integrated (production contract) laying hen operation. He stated that the family worked less for more net return with the latter. He also stated that production contracting was saving family farms in Mercer County. Several surveys of broiler and hog producers have indicated considerable satisfaction with contractors (see Tweeten and Flora 2001, p. 25). Less satisfaction is expressed with economic returns, but contract producers seem to be more satisfied than independent producers.

Despite these advantages, populists condemn production contracting. Neil Harl (2001, p. 49) asserts that "without much doubt the greatest economic threat to farmers as independent entrepreneurs is the deadly combination of concentration [in agribusiness] and vertical integration." He worries that farmers "without meaningful competitive options" will become "serfs" (p. 45).

Production contracting saves labor, hence, fewer family farms are needed. In Ohio, only one person-hour per day is required to operate a typical one-thousand-hog finishing barn. Thus, one person can operate as many as eight barns and 2.5 generations per year for a total of twenty thousand finished hogs. (My father worked pretty hard to farrow-to-finish three hundred hogs per year when I was a youth.) With the huge labor productivity on contract farms, perhaps it is not surprising that production contract farms in the United States average five times the size of the average U.S. farm–yet are family farms relying mostly on family labor.

Thus, a principal impact of production contracts is to speed adoption of modern technology and realize economies of size in production and marketing. Gains have been passed to consumers. Farms increased in size and labor efficiency mainly in response to new technology and incessant demand by consumers for a product that meets their wants. Thus structural change likely would have occurred but belatedly in the absence of production contracts.

It appears that the populists' anger is more with scale of operations than with contracting. Large-scale livestock farming is efficient and drives down market prices for all. This creates problems for small, less-efficient independent operators. Other problems may be odors, flies, water quality, and road damage. Improved management and technologies are diminishing these problems. Experts cited in chapter 6 contend that if environmental regulations are put in place and enforced for all farms, then large farms will continue to produce more cheaply per unit than small farms (see also Tweeten and Flora 2001).

The U.S. Senate in early 2002 passed a populist proposal to prohibit beef and pork packers from owning, feeding, or controlling livestock more than 14 days before slaughter. The term "controlling" in the measure seemed to rule out marketing contracts as well as production contracts. Marketing contracts, unlike production contracts, leave ownership in the hands of feeders but are widely used by packers to assure packers of delivery of the hogs and cattle for slaughter with specified genetics, feeding, and price needed to utilize slaughter capacity efficiently while meeting customers' demands for quality meat. The producer is assured of a market at the specified forward price.

The Senate proposal was heavily criticized on several grounds by leading livestock economists (Fenz et al. 2002, pp. 1-10):

- The proposal if enacted would deny packers opportunities to deliver the right product at the right time and place at the right price to consumers. Unstable delivery time and poor quality cattle supplied without the benefit of forward contracts substantially raise packer costs and threatens their viability especially in regions of marginal competitiveness.

- Consumers could end up paying more for a lower quality of meat. Studies indicate that higher processing costs are passed to consumers (Persaud and Tweeten 2002).
- Livestock producers would face greater risk without the benefit of forward contracts. Because of problems with meat quality in consumer markets and because of higher meat processing costs, farm prices and marketings might be lower.

Other problems could appear, such as less international competitiveness of U.S. pork and beef. It is apparent that the populist proposal designed to protect independent producers could make producers, consumers, packers, and the nation worse off. Fortunately, the Senate-House conference committee excluded the measure from the 2002 farm bill.

Several additional observations are listed below in concluding this subsection:

- More comprehensive public reporting of production contract terms is needed to ascertain, for example, whether contractors take advantage of growers who have sunk building costs.
- The nation's antitrust and environmental laws need to be reviewed, updated if necessary, and enforced.
- More comprehensive monitoring of environmental impacts by farms of all sizes is needed along with research to better address waste disposal problems that plague livestock farmers.
- Communities need to recognize that because of economies of size apparent in most livestock operations their option is not small farms versus large farms. For the most part, nonhobby small farms will be unable to compete and will go out of business. Except for hobby farms and some highly innovative niche and lifestyle producers, the option facing rural communities is large dairy and livestock feeding farms or no livestock farms.
- Core farm structure and environmental laws need to be national in scope although with allowance for unique population density and aridity conditions by region. State A passing laws to keep production contracting, corporate, or large-scale farms out may damage local family farms. That is, large-scale or production contract farms may go to another state where their output and efficiency will drive down nationwide livestock prices. Thus, the family farmers in state A may be driven out of business and not have recourse to the institutions and earnings they need to survive.

Contracting firms, veterinarians, processors, feed dealers, and other infrastructure are likely to follow the large-scale farms out of state. Remaining small farms could be hurt by less infrastructure. And food processors need competitively priced farm commodity inputs for their plants. South Dakota, which has the most restrictive of the anticorporate farming laws among the eight states with such laws, has seen many opportunities for economic development bypass the state.

**Myth 10: Agribusiness exploits farmers.**

The most consistent populist theme is that farmers are the victims of corporate agribusiness greed. The incessant quest for profit in the firms that supply inputs to farmers and in the firms that process, transport, store, and sell farm and food has destroyed the family farm according to populist thinking.

The populist literature is boundless, and it is possible here to catch only some of the flavor. William Heffernan (1999, pp. 12, 13) expressed concern for farmers over concentration of market power in clusters of agribusiness firms, and predicted that

> as the food chain clusters form, with major management decisions made by a small core of firm executives, there is little room left in the global food system for independent farmers. ...If the number of farmers is reduced to about 25,000 in the next decade, there will be many farm families who will be involuntarily removed from their land.

Heffernan's presumption of only 25,000 farms remaining in a decade is premature. The number of farms fell from 1,925,700 in census year 1992 to 1,911,859 in 1997, the latest census year, or at a rate of only 0.14 percent per year (NASS 1999, p. 10). At this rate, 3,096 years instead of Heffernan's predicted ten years will be required to reach 25,000 farms. The rate of loss of farms, in fact, has slowed as agribusiness concentration has grown. That is a chance happening, but would not have occurred if concentration drove out many farms.

Earlier, Maurice Dingman, Catholic Bishop of Des Moines, Iowa, shared his insights into the source of and solution to farm difficulties (1986, pp. 3-9):

> The [farm] Problem is a value crisis. There has been a shift from agriculture to agribusiness. ...The concepts implicit in the [1962 study *An Adaptive Program for Agriculture*] were clear; the values, lives, and communities of small farmers on the land were of less worth than technological innovations and opportunities for capital investment on the part of corporations and banks.
>
> ...To solve the economic crisis in rural America family farmers must unite for effective legislation and instead of competing with each other at the market place, must unite their production collectively nationwide. ...If you were a Catholic and you rejected collective bargaining, would this be a matter of concern at your next Confession?...Is there an obligation to join the National Farmers Organization?

**Box 8.2**

Following Bishop Dingman's presentation, I overheard another economist at the conference quip to a companion that "Bishop Dingman should be renamed 'Bishop Dingbat.'" A few days later I received a copy of an article on the front page of *The Witness*, a Catholic newspaper published in Iowa, stating that "Professor Tweeten calls Bishop Dingman 'Bishop Dingbat.'" The article went on to state that I should spend purgatory trying to make a living on a 300-acre Iowa farm. After receiving several hate letters from irate readers, I protested to editors of *The Witness* that they had fingered the wrong man and had transgressed the eighth commandment regarding false witness. I received no apology from *The Witness,* but they allowed me to publish a short disclaimer. I have since confounded learned theologians with a query: Must I spend time in purgatory even if I don't believe in it?

**Box 8.3**

Emotion was apparent in Bishop Dingman's condemnation of *An Adaptive Program for Agriculture* cited earlier. That forward-looking policy statement from the Committee for Economic Development (CED) was published in 1962. The statement called for conversion of all commodity programs to direct, decoupled payments that would be reduced by equal amounts each year until they were phased out in ten years—a framework later adapted for the 1996 farm bill. Populist farmers reacted strongly against the report. Among other actions, they gathered at Sears stores in Kansas City and elsewhere, depositing their catalogs in protest of Sears, which was one of the organizations financing the nonpartisan CED.

An American Agriculture Movement activist and retired army officer from the Oklahoma Panhandle spoke at Oklahoma State University about 1980, advising his audience that he did not want to see his tax dollars going to a university with Luther Tweeten on the faculty. My transgression was to serve as advisor to CED in its new report on agricultural policy. That new report was bland and unthreatening, but I was being lumped with the earlier, notorious 1962 report that I had nothing to do with. The activist ranted against the Trilateral Commission, which allegedly was tied through interlocking directorates to the CED in a conspiracy. He said the conspiracy had three aims: (1) to destroy confidence in the U.S. federal government, (2) to infiltrate the military, and (3) to control the nation's food supply. I wrote a tongue-in-cheek letter to the activist saying it was apparent to me and others that he was an emissary of the Trilateral Commission. As evidence, I cited the fact that his entire speech was an attack on federal government policies, he had infiltrated the military as an army officer, and he was a member of the American Agriculture Movement with an avowed aim of controlling the food supply to obtain 100 percent of parity prices. The moral of this anecdote is that guilt by association is cheap, easy, and often senseless.

Bishop Dingman delivered these comments at a 1986 conference entitled *Is There a Moral Obligation to Save the Family Farm?* held at Iowa State University (Box 8.2). Signs of conflict were apparent in an early press release that headlined the conference as "Economists versus Moralists" with the moralists being clergypersons, academic philosophers, and everyone except economists. The clear implication was that economists were amoral if not immoral!

In chapter 1, I cited the Organization for Competitive Markets (OCM) formed in 1998 as an example of an organization with a generalized, emotional disdain for corporate agribusiness. It merges postmodern philosophy and agricultural populism. Whereas conventional agriculturists mostly subscribe to Descartes' Enlightenment philosophy that reason rather than emotion makes for sound decisions, OCM (March 2000, p. 1) contends that "emotion is an absolute necessity for reason" (see Box 8.3).

Inflammatory rhetoric from OCM or other populists demonizing agribusiness can lead some farmers who are on the edge emotionally because of financial or other stress to blame agribusiness for their personal difficulty. Farmers have murdered a number of agribusiness persons. How much did an atmosphere of scapegoating by populist groups contribute to such tragedies? No one knows. But unsubstantiated categorical blame of predatory behavior by the agribusiness sector for farmers' economic ills is not helpful.

At issue is whether the populist critique of agribusiness has merit. In evaluating that issue it is useful to divide agribusiness into (1) input supply and (2) product marketing sectors. Neither sector is perfectly competitive, that is, having so many firms that one firm can ignore the actions of other firms when changing price. Both sectors are oligopoly, characterized by a few sellers, or oligopsony, characterized by a few buyers.

First, consider the food-marketing sector that buys, transports, stores, processes, and finally retails farm products. Recent studies examine the behavior of marketing margins between farmers and consumers. Of concern is whether rising concentration of economic activity in a few firms changes marketing margins. Increasing concentration of marketing activities in fewer firms can reduce marketing margins by achieving economies of size to lower costs. Or increasing concentration can raise marketing margins by increasing marketing power of firms so they can pay farmers less and charge consumers more.

The finding from empirical studies is that the economies of size dominate market power (Azzam and Pegoulatos 1994; Azzam 1997; Persaud 2000). That is, increasing concentration of marketing activities in the hands of fewer firms has reduced marketing margins, other things equal. That is the good news; the bad news for farmers is that the reduced margin of marketing firms is passed to consumers and not to farmers.

This outcome is predicted by economic theory. Whether marketing firms act competitively or monopsonistically (as single buyers), they pay farmers only the marginal cost of production. In other words, farmers will be paid what it takes to supply the product whether the buyer is a small firm, large firm, concentrated-industry firm, or cooperative firm.

Compared to a competitive marketing sector of many firms, the marketing sector operating as a single seller to consumers would buy less and pay less for the raw product from farmers, but the purchase would be along the farm supply curve. If the single seller were also a single buyer of farm commodities, it would operate even farther down the farm supply curve–at a lower price and quantity. So does imperfect competition in the food marketing sector cause farmers to be lower on their supply curve? The answer is no.

Economists don't know enough about the behavior of oligopsony and oligopoly that characterizes the food-marketing sector to predict pricing and output, but know that firms so structured advertise and innovate mightily. The consequence is a food-marketing sector so successful in devising tasty foods and advertising them widely that Americans chronically overeat on average about 12 percent, and half of Americans are overweight (Finke and Tweeten 1996).

The conclusion is that American farmers selling to an imperfectly competitive marketing sector in all likelihood operate higher on the food supply curve. That is, they receive higher prices for a larger aggregate volume of raw materials than they would if they sold to a competitive marketing sector that would not advertise or innovate. The fact that approximately one-third of farm output is marketed through cooperatives provides further insurance that farmers are not exploited by the marketing sector. Furthermore, the marketing sector would hardly be wise to destroy the farm sector that provides the raw materials on which the marketing sector exists.

Now turning to the farm input supply sector, we have less rigorous research to draw upon. The pace of mergers has been rapid in part because of intense competition that has driven many of the great names in agricultural input supply out of business. In machinery manufacturers alone, gone are International Harvester, Oliver, Minneapolis-Moline, Allis Chalmers, Silver King, Ford, and others because they could not make enough profit. In recent years many high-tech firms providing genetically modified new plant varieties that might be expected to be profit leaders have merged, consolidated, or ceased operations for lack of profits. Among other constraints on excess profits, domestic and foreign farm input suppliers are eager to enter and expand in the large American market if profits beckon. They constitute what industrial organization economists call a "credible threat," encouraging competitive behavior in an imperfectly competitive industry.

If input supply firms are wielding market power to accrue excessive profits, then cooperatives should prosper along with private agribusiness firms. Producer cooperatives account for about one-third of the farm input supply market, reducing chances for exploitation of farmers. To better compete, a number of cooperatives have integrated vertically to operate in nearly all phases of farm input supply, contracting, product processing, and product marketing. They have not prospered in competition with private firms and, indeed, many would not survive without government help. Cooperatives have consolidated at a rapid pace in recent years to compete and survive. The largest farm cooperative, Farmland Industries, filed for bankruptcy in 2002.

I know of no empirical study indicating that anticompetitive agribusiness behavior is causing farms to get larger and fewer (Barse 1990; Leibenluft 1981). Farms as well as agribusiness firms are consolidating for the same reasons that all industries are consolidating. These reasons include availability of large, expensive, indivisible technological and human capital that reduces costs per unit of output. Costs per unit are reduced, however, only if that "lumpy" capital is spread over many units of output.

Firms are consolidating also to gain advantages of task specialization. That is, costs are lower and efficiency higher by having specialists in the respective fields of management, information systems, marketing, finance, and blue-collar activities. An all-purpose family member in a family firm performing each of these tasks will not do so very efficiently. Financing expensive research and development, advertising in national media, coping with risks, meeting government regulations (e.g., food safety and quality, environment), and obtaining access to national venture capital markets are also reasons to lower unit costs by expanding size through firm growth or consolidation.

Millions and sometimes billions of dollars are required to research and develop each new sophisticated technology through bioengineering, computer assisted design, and other means. A firm must have command over massive resources to afford the human, material, and technological capital essential for success. To survive, it must be able to engage in enough ventures so that the few successful ventures will offset the many failures.

In an atomistic (many firm) industry, each firm can charge only the marginal cost for its products. That means in postindustrial agribusiness where millions of dollars may be sunk in overhead before a successful product is introduced, competitive firms cannot survive. Market power through intellectual property rights or other means is required for firm survival. Farmers and the groups that represent them will have to learn how to live with that kind of world.

Overhead costs are incurred whether or not firms are operating. With high overhead relative to operating costs, a premium is placed on a reliable, steady supply of raw material to keep processing plants operating near capacity and overall costs per unit low. Firms can utilize production and marketing contracts to ensure product to process and thereby keep total unit processing costs low. Koontz (2000, p. 5) states, "I would argue that little contract production has emerged because of power. It has emerged to produce a product more consistent with low-cost processing systems and consumer wants."

To summarize, farming has been far more influenced by favorable performance of agribusiness bringing increased productivity than by unfavorable conduct bringing high farm input prices or low commodity prices through market power. The major source of decline in number and increase in size of family farms has been technology, especially adoption of laborsaving farm machinery. Such technology is the result of ingenuity and is not the result of predatory behavior or subpar performance of agribusiness. Scale-influencing technologies would have caused losses in commercial farm numbers even if farm prices would have been much higher or production contracts had never been devised. Productivity gains have brought massive national income benefits to society as a whole and hence to farmers in the long run because farm income per capita has trended toward national income per capita. A major source of the spectacular rise in farm household income since the 1930s has been laborsaving technology that has freed farmers to operate larger units and work off farms.

## CONCLUSIONS

Agricultural populism seems to be another name for bad economics. The ten myths of populism examined in this chapter by no means exhaust the list. But to consider more would exhaust the reader.

An end to agribusiness concentration and the international corporate business and finance "conspiracy" could not solve the problem of low resource returns chronically experienced by most farmers. The reason is that most farms are too small and/or poorly managed to earn resource returns comparable to those earned elsewhere. (Most farm output is produced by farms well enough managed to earn a positive return, however.) Other sectors do not experience such chronic low returns to a majority of firms because they do not have the tradition of their occupation being a hobby financed by government programs.

Having mostly high-cost farms does not necessarily imply economic inefficiency. Today's farmers move readily between farm and off-farm employment. They have freely chosen to reside on high-cost farms and pay for that hobby with off-farm income.

The essential point is that populist solutions to end the alleged agribusiness conspiracy will not work because they do not address root causes. Hence, the populist feelings of exploitation by agribusiness will not go away even if by some magical policy all agribusiness were to become perfectly competitive. And populist calls for supply management and high commodity price supports, if implemented, would speed farm consolidation and reduce family farm numbers, making the patient worse.

Small farms have accommodated to market realities by covering their farm losses with off-farm income and large farms producing most of the nation's food would perform well in a market economy. If the government decides that the vulnerable midsized farms need help, it could assist them at a fraction of current program cost with measures specifically targeting them while current commodity programs are ended.

One cannot help but be struck by the stark contrast between vilification of markets and agribusiness industries by populists and the absence of evidence justifying such vilification through numerous in-depth economic studies. Nonetheless, studies of agribusiness market structure (size, number, concentration of firms), conduct (predatory, exclusionary, price-fixing behavior), and performance (efficiency, innovation, etc.) need to continue. U.S. antitrust laws need to be reviewed, revised if necessary, and enforced.

The most innovative and productive agriculture in the world rests on the three pillars: able farmers, science, and agribusiness. Each of these has been a critical component, and neither of the other two could have succeeded without the excellent performance of the agribusiness industry. The first order of public policy is to do nothing to make the "patient" worse with ill-conceived populist nostrums that emerge when one of the three pillars of agriculture turns against another.

## REFERENCES

Azzam, A. "Measuring Market Power and Cost-Efficiency Effects of Industrial Concentration." *Journal of Industrial Organization* 45(1997): 377-86.

Azzam, A., and E. Pagoulatos. "Testing Oligopolistic and Oligopsonistic Behavior: An Application to the U.S. Meat-Packing Industry." *Journal of Agricultural Economics* 41 (1990): 362-70.

Barse, J., ed. *Seven Farm Input Industries*. Agricultural Economic Report Number 635. Washington, DC: Economic Research Service, USDA, September 1990.

Bentley, S., P. Barlett, F. Leistritz, S. Murdock, W. Saupe, D. Albrecht, B. Ekstrom, R. Hamm, A. Leholm, R. Rathge, and J. Wanzek. *Involuntary Exits from Farming: Evidence from Four Studies*. Agricultural Economic Report No.625. Washington, DC: Economic Research Service, U.S. Department of Agriculture, November 1989.

Council of Economic Advisors. *Economic Report of the President*. Washington, DC: U.S. Government Printing Office, 1987, 2000, 2001 and other issues.

Dingman, M. "What Does Christianity Have to Do with the Farm Crisis?" Proceedings of conference *Is There a Moral Obligation to Save the Family Farm?* Ames: Religious Studies Program, Iowa State University, 1986.

Ehmke, T. "Kansas' Most Wanted." *Top Producer*, January 2002, pp. 15-27.

Fenz, D., G. Grimes, M. Hayenga, S. Koonz, J. Lawrence, W. Purcell, T. Schroeder, and C. Ward. *Comments on Economic Impacts of Proposed Legislation to Prohibit Beef and Pork Ownership, Feeding, and Control of Livestock*. Ames: Iowa State University Beef Center, January 14, 2002.

Finke, M., and L. Tweeten. "Economic Impact of Proper Diets on Farm and Marketing Resources." *Journal of Agribusiness* 12 (1996): 201-07.

Goodwin, B. "Instability and Risk in U.S. Agriculture." *Journal of Agribusiness* 18 (March 2000): 71-79.

Harl, N. "The Structural Transformation of the Agribusiness Sector." *Fixing the Farm Bill*, edited by J. Schnittker and N. Harl. Proceedings of conference held in Washington, DC. Ames: Department of Economics, Iowa State University, 2001.

Heffernan, W. *Consolidation in the Food and Agriculture System*. Report to the National Farmers Union. Columbia: Department of Rural Sociology, University of Missouri, 1999.

Hopkins, J., and M. Morehart. "An Empirical Analysis of the Farm Problem: Comparability in Rates of Return." In *Agricultural Policy for the 21st Century*, edited by L. Tweeten and S. Thompson.. Ames: Iowa State Press, 2002.

Koontz, S. *Concentration, Competition, and Industry Structure in Agriculture*. Testimony at Agricultural Concentration and Competition Hearing, April 27, 2000.Washington, DC: Committee on Agriculture, Nutrition, and Forestry, U.S. Senate, 2000.

Langemeier, M., and G. Patrick. "Farmers' Propensity to Consume: An Application to Illinois Grain Farmers." *American Journal of Agricultural Economics* 72 (1990): 309-25.

Leibenluft, R. *Competition in Farm Inputs: An Examination of Four Industries*. Washington, DC: Federal Trade Commission, February 1981.

NASS, U.S. Department of Agriculture. *1997 Census of Agriculture*. "United States Summary and State Data." AC97-A-51. Washington, DC: National Agricultural Statistics Service, March 1999.

NORM (National Organization for Raw Materials). "Welcome to NORM." http://www.normeconomics.org/welcome.html, 2002.

OCM. *Newsletter*. Lincoln, NE: Organization for Competitive Markets, March 2001 and other issues.

Perry, J., D. Schreiner, and L. Tweeten. *Analysis of the Characteristics of Farmers Who Have Curtailed or Ceased Farming in Oklahoma*. Research Report P-919. Stillwater: Agricultural Experiment Station, Oklahoma State University, 1991.

Persaud, S. *Investigating Market Power and Asymmetries in the Retail-to-Food and Farm-to-Retail Price Transmission Effects*. PhD Dissertation. Columbus: Department of Agricultural, Environmental, and Development Economics, Ohio State University, 2000.

Persaud, S., and L. Tweeten. "Impact of Agribusiness Market Power on Farmers." In *Agricultural Policy for the 21st Century,* edited by L. Tweeten and S. Thompson.. Ames: Iowa State Press, 2002.

Sanford, S., L. Tweeten, C. Rogers, and I. Russell. *Origins, Current Situation, and Future Plans of Farmers in East Central Oklahoma*. Oklahoma Agricultural Experiment Station Research Report P-861. Stillwater: Agricultural Experiment Station, Oklahoma State University, November 1984.

Tweeten, L. *Farm Policy Analysis*. Boulder, CO: Westview Press, 1989.

Tweeten, L. "Is It Time to Phase Out Commodity Programs?" In *Countdown to 1995: Perspectives for a New Farm Bill.* Anderson Publication ESO 2122. Columbus: Department of Agricultural Economics, Ohio State University, 1994.

Tweeten, L. "Farm Commodity Programs: Essential Safety Net or Corporate Welfare?" In *Agricultural Policy for the 21st Century,* edited by L. Tweeten and S. Thompson. Ames: Iowa State Press, 2002.

Tweeten, L., and C. Flora. *Vertical Coordination of Agriculture.* Task Force Report 137. Ames, IA: Council for Agricultural Science and Technology, March 2001.

Tweeten, L., and C. Zulauf. "Farm Succession: Who Will Farm in the Twenty-First Century?" In *Research in Domestic and International Agribusiness Management,* edited by Ray Goldberg. Vol. 11. Greenwich, CN: JAI Press, Inc., 1995.

U.S. Department of Agriculture. *The Farm Income Situation*. FIS-179. Washington, DC: Agricultural Marketing Service, USDA, July 1960.

U.S. Department of Agriculture. *Changes in Farm Production and Efficiency, 1978*. Statistical Bulletin No. 628. Washington, DC: Economics, Statistics, and Cooperative Service, USDA, 1980.
U.S. Department of Agriculture. *Agricultural Income and Finance*. AIS-72 Sept. 1999, AIS-74 Feb. 2000; and AIS-75 Sept. 2000. Washington, DC: Economic Research Service, USDA.
U.S. Department of Agriculture. *Agricultural Outlook*. Washington, DC: Economic Research Service, USDA, March 2001, October 2001.
Walters, C., Jr. *Unforgiven, the Biography of an Idea*. Kansas City: Economics Library in Cooperation with Citizens' Congress for Private Enterprise, 1971.
Wright, B. "Goals and Realities for Farm Policy." *Agricultural Policy Reform in the United States,* edited by Daniel Sumner. Washington, DC: AEI Press, 1995.

# 9

# Farm Organizations, Protest, and Populism

Chapters 7 and 8 outlined the misinformation and costs generated by populist agriculture. This chapter sheds light on another dimension of populist agriculture–its capacity for organization, protest, and, at times, violence.

Americans are impressed by the tranquility of the rural scene. Many nonfarmers may have the impression that all farmers are committed to goals and values compatible with democracy, individualism, independence, and the "orderly" American way of life. However, agricultural history is punctuated with discordant notes–by attempts on the part of the farmers to organize and by dramatic, sometimes violent, expressions of protest motivated by populist ideology.

This chapter enumerates various instances in which farmers took steps outside the market to deal with pressing economic issues integral to the unfolding commercialization of agriculture, but which at the time they did not understand. Many of the grievances to which farmers reacted were real, many imagined. Whatever the source of the grievances, they contain real lessons for farm policy. The roots of several farm organizations lie in populist ideology and protest movements. The origin and progress of the major farm organizations are also discussed, including the American Farm Bureau Federation, which was one of the few general farm organizations not arising from populist ideology and protest movements.

Much of early American agriculture was subsistence farming. There were few protest movements until farmers dealt with commercial outsiders they could protest against. The outsiders were primarily firms and institutions dealt with by farmers in their buying and selling activities. As the economy became more complex, and as business and government grew, farmers not only dealt with the local general store and its owner but with the local grain elevator and thus indirectly with the great transportation and marketing industries. Government became larger as it was called upon to perform a wide range of functions. The bigness, complexity, and remoteness of big government and industry often seemed inscrutable and even sinister to farmers. Whether their complaints were justified or not, farmers in fact found the nonfarm sector a ready scapegoat for all kinds of economic ills.

## EARLY PROTEST MOVEMENTS

Colonists learned tobacco culture from the Indians and planted the crop as early as 1610. By 1639 they were shipping 1.5 million pounds annually to the English markets. About 1630, prices of tobacco fell sharply. In response to complaints of farmers, legislative price-fixing and acreage controls were attempted but failed. Farmers organized collective bargaining groups to curtail production by voluntary agreements among themselves but were unsuccessful in stabilizing production or markets. In 1682, rioters engaged in plant cutting and destruction of tobacco, and their efforts were said to have improved the price in 1683 (Taylor 1953, p. 21).

Numerous additional attempts were made by tobacco planters' clubs, by merchant organizations, and by legislation to stabilize tobacco prices or control production prior to 1750. These efforts of farmers to bargain collectively–and the efforts of the Virginia government to stabilize tobacco prices by destroying the already processed crop and by paying farmers to plant soil-improving crops rather than tobacco–were of some economic success. That success was short lived, and probably had little long-run influence on tobacco prices, production, or farm income. However, tobacco "wars," often characterized by riots and destruction of tobacco by farmers, were the forerunners of the great farm protest movements that are a part of U.S. history.

### Shay's Rebellion

The Articles of Confederation did not give Congress control over currency and banking. Each state was individually responsible for its banking and currency, hence the economy was ill equipped to deal with the deflation in the 1780s that followed great inflation during the revolution. Taxes did not decline with prices. Debts incurred in periods of high prices were difficult or impossible to repay. Debtors were imprisoned and mortgages foreclosed. Most of the debtors were farmers; the creditors were merchants and bankers.

Debtors first called for settling by arbitration. They later called for induced inflation by issuance of currency so that debts could be repaid more easily. In New England farmers resorted to destruction of produce, market strikes, direct appeals, legislative intervention, and sometimes to armed revolt and violence to make their case heard. They burned the barns and haystacks of those who would not participate in the protest. Farmers carried their fight to the newspapers and the legislature. In Rhode Island they succeeded in passing a law that required merchants to accept paper money at face value in place of gold. Merchants responded by closing their doors and the farmers went on strike in 1786. The law was later revoked and farmers lost this fight for cheap money (Taylor 1953, p. 28).

Politicians and commercial interest said that the unfavorable economic conditions would be alleviated if farmers would just work hard, practice thrift, diversify production, and be more self-sufficient. These suggestions did not give the quick relief sought by farmers.

At the height of the rebellion in 1786, approximately five thousand persons, mostly farmers, were in open rebellion in Massachusetts. Daniel Shay's rebels were prominent. According to Taylor (1953, p. 36):

> The outcome of Shay's Rebellion was defeat after a number of actual clashes of arms between the rebels and the militia. In March 1787, the leaders were tried and 14 of them convicted of treason and sentenced to death. Governor Bowdoin granted them a reprieve for a few weeks, and Governor Hancock, his successor, pardoned all of them. Thus ended the first formidable insurrection ever to occur in the history of our new nation. Although debtors of all kinds were among Shay's troops, it was primarily a farmers' revolt against debts, taxes, and low prices.

Protesting farmers were not strong enough to win their demands. As in so many subsequent protests, direct action accomplished little more than to focus attention on the plight of farmers. The country was ill equipped by fiscal-monetary theory and nonfarm resources to legislate favorable economic conditions for farmers, but it did grant the protestors some concessions and treated them with a degree of leniency that prevented an even more serious long-term alienation.

## FIRST FARM ORGANIZATIONS

The New England Association of Farmers, Mechanics, and Other Workingmen is believed to be the first "farmer" political organization (Taylor 1953, pp. 77-79). Farmers were concerned about low prices, the practice of imprisonment for debt, and compulsory military drill. Many of their sons and daughters were employed in industry, hence farmers were concerned about low wages, long factory hours, and unemployment. The New England group was established about 1830, and many similar groups sprang up in the Northeast and in the Ohio Valley.

Numerous farm groups and societies were organized from 1830 to 1870. Many were local groups that had comparatively few political interests. The Patrons of Husbandry emerged as the first farm organization whose weight was felt in economic matters and farm policy formation.

## PATRONS OF HUSBANDRY

Farming was not sufficiently commercial, grievances were too diversified, and overriding national issues such as slavery were too pressing for farm discontent to express itself widely until after the Civil War. The time for organization and protest was ripe. The index of prices received by farmers fell from 119 in 1860 to 93 in 1872 (1910-1914 = 100). Prices continued downward through the 1870s and reached an index of 67 in 1879. Nearly all farmers were antagonistic to the "monopolistic" corporations that were said to ask too much for what farmers bought and pay too little for what farmers sold. They were antagonistic to the railroads, which constantly engaged in rate discrimination and rate wars, and to the bankers, who they thought charged too much interest, foreclosed on mortgages too readily, and created tight money, deflation, and depression (Tweeten 1979, ch. 3).

Never in American history had farm prices been depressed for so long a time as they were in the 30 years following the Civil War, nor has there ever been a period in which so large a percentage of American farmers rose in protest against the conditions in which they found themselves (Taylor 1953, p. 91). Farmers, Taylor writes,

> welcomed and sought opportunities to produce for the market, but they did not seem to understand that a price and market economy requires finance and credit institutions, transportation and other shipping agencies, manufacturers, middlemen, and even additional government services, and thus more taxes. They, therefore, protested against these and supported public men who were, so to speak, anti-industry, anti-bank, and anti-government. (p. 42)

In 1866, Oliver Kelly, as an employee of the recently established Department of Agriculture, toured the South to reestablish statistical reporting after the Civil War. He was a former Minnesota farmer who, because of drought conditions, had sought work with the U.S. Department of Agriculture. The deprivation, illiteracy, and lack of social life in the South made a deep impression on him. He resolved to do something about it and, with the assistance of a small group of concerned individuals, founded in 1867 the National Order of Patrons of Husbandry, commonly called the Grange. Because Kelly and some of the other founders were Masons, the new farm organization was established as a secret social and educational order with a special ritual.

Membership increased slowly at first. By the end of 1871 less than 200 local groups had been formed, mostly in Minnesota and Iowa. But the benign fraternal order was on the verge of becoming the national instrument to express farmer discontent. The Grange provided the organization, discontented farmers the membership. In 1872, 1,150 new locals were formed, more than half of them in Iowa. The number of locals numbered 21,697 by the end of 1874. The national leaders were reluctant to make the Grange the tool of agrarian discontent, but the forces of discontent were too strong to oppose. Sharply rising membership, which totaled 268,000 in 1874 and 858,000 in 1875, clearly established the Grange as a political force to be reckoned with. R. L. Tontz (1964) estimated its membership to be 451,000 families in 1875 (table 9.1). The Grange led the farmers' successful fight for public policy reform in the 1870s.

**The Granger Laws**

Many of the Grange's political activities were aimed at regulation of railroads. Regulations included these features: fixing of maximum rates by a public commission, prohibiting rate discrimination between long and short hauls (the total charge of a short haul could not exceed that of a long haul), forbidding consolidation of parallel rail lines that would reduce competition, and prohibiting the granting of free passes by railroads to public officials.

**Table 9.1. U.S. Membership in General Farm Organizations, Selected Years, 1875 to 2001**

| Year | Grange | Farmers Alliance | Farmers Union | Farm Bureau | Total |
|---|---|---|---|---|---|
| Thousands of families | | | | | |
| 1875 | 451 | | | | 451 |
| 1880 | 65 | 39 | | | 105 |
| 1885 | — | 232 | | | — |
| 1890 | 71 | 1,053 | | | 1,124 |
| 1892 | — | 61 | | | — |
| 1893 | — | 26 | | | — |
| 1900 | 99 | — | | | 99 |
| 1908 | — | — | 135 | | — |
| 1910 | 224 | — | 117 | | 340 |
| 1912 | — | — | 116 | | — |
| 1914 | — | — | 103 | | — |
| 1916 | — | — | 107 | | — |
| 1918 | — | — | 129 | | — |
| 1920 | 231 | — | — | 317 | 678 |
| 1925 | — | — | — | 314 | — |
| 1930 | 316 | — | — | 321 | 716 |
| 1933 | — | — | 78 | 163 | — |
| 1935 | — | — | — | 281 | — |
| 1938 | — | — | 84 | — | — |
| 1940 | 337 | — | — | 444 | 866 |
| 1945 | — | — | — | 986 | — |
| 1950 | 443 | — | — | 1,450 | 2,109 |
| 1953 | — | — | 216 | 1,623 | — |
| 1956 | — | — | 278 | — | — |
| 1959 | 405 | — | — | 1,602 | — |
| 1960 | 394 | — | — | 1,601 | 2,273 |
| 2001 | 300 | — | 300 | 5,101 | 5,701 |

*Source:* Tweeten (1979) and private communication with organization officials for year 2001. Dashes indicate years for which data were not available.

The Grange's goal of public regulation of rail rates was realized with enactment of the so-called Granger Laws. The Illinois legislature in 1870 took the lead under the urging of the Grange supporters and established a number of constitutional directives to control railroad rates and warehouses. The following year the legislature passed laws implementing the directives. It also required annual reporting by railroad companies of assets, debts, monthly earnings, and expenses. The courts, however, reacted adversely, so a new act, more carefully formulated, was passed by the

Illinois legislature in 1873. Among other features it provided penalties for rate discrimination. The state supreme court upheld the constitutionality of this act in 1880. "Granger cases," such as *Munn vs. Illinois* in 1876, established the right of government to regulate public utilities.

Other states passed laws, often poorly formulated, to regulate railroads. These laws formed a discordant system of regulation. Mercifully, they were short-lived. The U.S. Supreme Court held that states could not regulate interstate commerce in the historic Wabash case of 1886 (Davis et al. 1965, p. 391). The following year, the federal government established a system to regulate freight and passenger service with passage of the Interstate Commerce Act of 1887.

The Grange was also active in national politics. It opposed the single gold standard and redeemable greenbacks. The organization was a major force behind raising the Department of Agriculture to cabinet status and was influential in providing for dissemination of agricultural literature, for agricultural fairs, and for teaching of agriculture in the public schools. It was influential in establishing agricultural experiment stations in 1887 and in obtaining rural mail delivery. It worked for elimination of fraud and adulteration in food processing, against bribery and corruption of public officials, and for resource conservation, women's suffrage, and ballot reform. Many of these causes were not immediately successful but later were nationally accepted and enacted into law. Many regulations were dismantled in the 1970s and subsequent decades. The fraternal aspects of the Grange, though submerged, undoubtedly added to the quality of rural life.

**Economic Program**

Numerous activities were undertaken of a rather direct economic nature. The Grange engaged in cooperative buying and selling, manufacturing, and banking and even engaged in holding actions to achieve economic objectives. Mutual fire insurance companies were organized in every Grange state. In 1871 members began buying household and farm supplies cooperatively. The operation of cooperative stores organized by county or district Granges was widely practiced. Members would sometimes pool their production and bargain collectively with merchants to receive a higher price for produce. Most of these efforts were without production controls, but "in one instance at least, an attempt was made by the Illinois Grangers in 1878 to influence the price of hogs by withholding them from the market" (Taylor 1953, p. 157).

Orders of many local Granges were pooled and state agents bargained directly with manufacturers for farm supplies. Sometimes agents were sent or were hired abroad to handle exports of U.S. farm products such as wheat and cotton. Local Grange cooperatives often handled the buying and storage of wheat. Nearly every phase of "agribusiness" was attempted, including manufacturing of farm machinery. Many of the cooperative elevators, shipping associations, and fire insurance companies were successful (Benedict 1953, p. 96). However, a large number of business activities failed from too much idealism and too little previous experience, capital, and business acumen. Also, unwarranted expectations, intense political involvement, and competition from private firms figured in the demise of Grange business interests. Nevertheless, the Grange provided useful experience and a foundation for business activities of the current farm organizations–including the National Grange of today.

The legislative successes of the Grange in the 1870s are a matter of history. Its accomplishments were impressive, and its ideas are still apparent in current farm organizations. But "it had grown too fast, it had attempted too much, and the methods of organization used were better suited to rousing enthusiasm than to maintaining it" (Benedict 1953, p. 104).

Grange membership peaked at 858,000 individuals (451,000 families) in 1875, only seven years after its founding, but dropped rapidly to 124,000 by 1880. It fell slightly lower but began to recover and by 1900 approached 200,000 members (Benedict 1953, p. 104). Emphasis on social and educational activities and a few sound economic activities coupled with deemphasis of politics gave considerable stability to the group in the twentieth century.

Grange membership was 400,000 individuals in year 2001, and these individuals in many cases represented families (table 9.1). Less than 10 percent of that membership was farmers. Thus, the Grange follows a pattern characteristic of farm populist protest movements that maintain long-term viability. The pattern is first to gain a following by promoting populist causes in protest movements. Attempts at food withholding actions (strikes) inevitably fail because farmers are too independent and scattered. When zeal for such actions wanes, the organization turns to political activities to obtain benefits from Washington. When individual members discover they can be free riders (obtaining government benefits without being a member), membership must be sustained by providing low-cost input supply and product marketing services to members. The most successful of these business services is insurance. And organizations eventually learn that with declining numbers of farmers, the best market for insurance and membership is nonfarmers.

## THE FARMERS ALLIANCE

The name "Farmers Alliance" was a popular one and was used by a number of independent farm organizations that began about 1873. Again, the fuel for the movement was farmers' discontent. The index of prices received by farmers stood at 95 in 1882 and fell steadily to 67 in 1886 (1910-1914 = 100). It remained low for several years thereafter. Populism was reenergized. Farmers viewed themselves as a class oppressed by railroad, banks, Wall Street monopolies, government, and the urban-industrial process.

### The Southern Farmers Alliance

The first Farmers Alliance is believed to have been formed in Texas in 1873 as an "anti-horse-thief" and "anti-land-grab" farmers' organization (Taylor 1953, p. 194). In other areas, the focus of farmers' discontent was government grants to railroads of land already occupied by settlers. One of the accomplishments of the Alliance in Kansas was to win court approval of preemption rights to land operated by squatters and to set aside the claims of railroads to preempted land (Benedict 1953, p. 106).

The Agricultural Wheel began in Arkansas as a local debating and discussion group called the Wattensas Farmers Club in 1882. It began innocuously enough. Like other farm clubs, it attempted to improve its members in the theory and practice of agriculture and to demonstrate ways to improve rural living. Its humble beginnings hardly foretold its later ambitious objectives: "action in concert with all labor unions or organizations of laborers" to secure legislation beneficial to farmers (Taylor 1953, pp. 203, 204).

At the time of a joint meeting with other farm organizations in 1887, total Wheel membership was reported to be 500,000 though it was probably not that high. Membership was mainly located in the south central region but extended as far north as Wisconsin. The Wheel had earlier joined the Brothers of Freedom, another Arkansas group that included not only farmers but also industrial laborers and townspeople. The Wheel opposed mortgages on livestock and crops, "soulless" corporations, and national banks. It proposed reducing crop acreage in order to lower production and raise farm prices (Benedict 1953, p. 106). It favored a graduated income tax and low tariffs and was much involved in politics.

The Louisiana Farmers Union originated following a discussion between 12 men cleaning a graveyard in 1880 (Taylor 1953, p. 200). By 1887, it had grown to 10,000 members. Meanwhile the Texas Farmers Alliance had grown to approximately 100,000 members. The two organizations merged in 1887. The North Carolina Farmers Association early in 1888 joined this combination. Later that year, this group joined with the Agricultural Wheel to form the Farmers and Laborers Union. The name was changed to National Farmers Alliance and Industrial Union in 1889 when the Laborers Union of Kansas joined the organization. The amalgamation is sometimes called the Southern Farmers Alliance. The Southern Alliance claimed membership of up to 3 million in 1890, but less than 250,000 supported the organization financially.

**The Northern or National Farmers Alliance**

A northern National Alliance was organized at Chicago in 1880. The organization, which stood for protection of farmers from the "tyranny of monopoly" and the "encroachments of concentrated capital," gained membership in the northwest central states totaling perhaps 400,000 in 1887. Membership was especially large in the Plains states, where low prices and hard times had increased farmers' discontent.

Numerous other farm groups were closely associated with the Farmers Alliances. The Farmers Mutual Benefit Association arose in Illinois in 1882 as an effort to improve wheat marketing. Membership of approximately 150,000 extended from Kansas to North Carolina but was concentrated mainly in the Midwest. Other groups such as the Alliance of Colored Farmers of Texas, Farmers Congress, Farmers League, and Patrons of Industry were closely associated with the Alliance movement, and most had representatives at an 1889 meeting in St. Louis to unite the Northern and Southern Farmers Alliances. The groups agreed on common objectives but failed to consolidate. The third-party idea seemed to divert attention from union (Benedict 1953, p. 109).

**Activities of the Farmers Alliances**

The Grange and the Farmers Alliances had many parallels. The Grange began as a social and fraternal group but was transformed by agrarian unrest in the 1870s. The Alliance movement–begun in widely separated localities as a discussion society and anti-horse-thief association (and for other reasons) –was radically changed in character by agrarian discontent in the 1880s. Both the Grange and Alliance movements were satisfied at first with comparatively mild measures to treat economic ills. The Alliance moved against middlemen by wholesale purchasing and manufacturing of farm supplies. Cooperative selling, insurance of many types, and credit also were provided. Many Alliance business ventures failed due to poor management, inadequate financing, or astute (sometimes unfair) competition from private firms (Taylor 1953, p. 235).

The "subtreasury plan" of the Southern Alliance is significant because, though never enacted into law, it was similar to the nonrecourse loan used extensively by the federal government to support farm prices since the 1930s. The plan was to establish a branch of the U.S. Treasury in every agricultural county. The owners of farm products would deposit their commodities in warehouses and receive legal tender for up to 80 percent of the market value of the commodity. The farmer could redeem the commodities and sell them as he chose by repaying the advance, plus carrying charges. If the farmer did not redeem the commodities in twelve months, they would be sold at public auction and the money used to cover operation of the system. The plan was designed to expand the currency and to remove the necessity to dump farm products on a sagging market (Taylor 1953, p. 244).

The Alliances were impatient for economic reform and were dissatisfied with the progress of local initiative and cooperatives, so they turned increasingly to redress in the political arena. This culminated in the joint effort with the Knights of Labor and other groups to support the Populist Party.

**Southern Alliance Submerged In the Populist Party**

The Alliances' militancy was pulling them more heavily into politics. They enjoyed substantial success in Kansas in 1890. According to Taylor (1953, p. 277), in 1892 when "in Omaha the Populist Party became a blazing reality, the Southern Alliance was literally buried." The National Peoples Party nominated John Weaver for president. Its platform called for free coinage of silver, a graduated income tax, return of railroad lands to the people, popular election of U.S. senators, secret ballot, an eight-hour workday, public ownership of railroads, and an effective civil service.

The issue that dominated all others was free silver. This was an agrarian issue to the extent that farmers hoped to gain from freer coinage and the easier credit that was thought to attend it. The Peoples Party did not win a majority of the legislative seats in any state, but it did poll many votes and was not completely discouraged. Weaver received 1.0 million votes, Cleveland 5.6 million, and Harrison 5.2 million (Benedict 1953, p. 111).

Populists and Democrats joined in 1896 to nominate William Jennings Bryan for president. He was defeated in the general election. The farmers' movement had been so thoroughly identified with politics that the setback virtually spelled the end of the Alliances and the third-party idea. One view is that they would have become extinct even if Bryan had won because the leadership had either died or been separated from the membership at the grassroots level (Taylor 1953, p. 323). Also, farm economic conditions were beginning gradually to improve in the last decade of the century, and farmers were finding it more difficult to unite behind a common cause.

The massive Alliance movement, which involved perhaps four million farmers, accomplished several important objectives. First, it gave the farmer a sense of worth and a consciousness of class. It reaffirmed as had the Grange earlier, that farmers could exert considerable political and economic power if pressed sufficiently by real or alleged wrongs caused by business and other groups. And the Alliance contributed to the experience accumulated earlier by the Grange, showing the need for sound business management and the danger of too heavy involvement in politics. The Alliance provided some of the groundwork for later, more successful farm organizations and their public policies.

## NATIONAL FARMERS UNION

Isaac Newton Gresham, a newspaperman and former member of the Farmers Alliance, founded the Farmers Educational and Cooperative Union in Texas. Gresham and nine of his friends received a state charter in 1902.

The Union absorbed former Alliance and Populist Party members as well as others and grew to an estimated 50,000 members in 1904. It was plagued by early strife, engendered by rather sizable membership fees coupled with loose accounting procedures and by the issue of whether only farmers could be officers. By 1905 the organization had membership in numerous states of the South. By 1908, membership totaled 135,000 farm families in twenty-three states (table 9.1). Membership after 1912 shifted westward, and the wheat states by 1920 were the major source of Union support. Unlike the old Alliance, the National Farmers Union grew little from amalgamations. However, it increased its membership in the Midwest by absorbing in 1907 a Farmers Union that previously had been formed by combining in Illinois the Farmers Social and Economic Union, the Farmers Relief Association, and the Farmers Mutual Benefit Association.

Whereas the Alliance had focused on politics to attack problems caused by railroads, monopoly, and deflation, the Union emphasized the business approach to problems of buying, selling, middlemen, and credit. The Union never placed a candidate for election or identified closely with one party in its early years. It avoided the political mistakes of the Alliance and the Grange. The Union pressed its political causes on conventions, platforms, and candidates, but did not tie its existence to any one political party or issue. In the 1930s it was involved in the Farmers Holiday Association, to be discussed later.

The charter of the first Farmers Union group clearly stated that the purpose of the organization was to aid farmers in marketing. After contracting with the cotton ginners in the county, in 1903 the first local claimed success in raising cotton prices. The Farmers Union in 1904 sponsored a movement to hold one bale of cotton out of five off the market and to sell the other four slowly. It took credit for raising cotton prices. In 1907 it attempted to limit production and control cotton prices by plowing down 10 percent of the cotton crop and holding some of the harvested crop off the market (Benedict 1953, p. 134). Warehouses were built by the Union to facilitate holding and marketing. In 1913, the Union had 1,600 warehouses in cotton states and wheat states (Taylor 1953, p. 353).

The Farmers Union was much involved in business activity. It attempted to produce implements and fertilizers. Its enterprises even included coal mines and banks. Purchasing clubs bought in carload lots to reduce middlemen's commissions. Some locals negotiated a 10 percent discount for Union members from retail merchants.

Considerable success was achieved with wholesale and retail cooperatives. These dealt in livestock, grain, insurance, and other items. The local stores handled a large number of items for the farm and household. The Farmers Union was the first organization to be really successful in business activities. This success gave continuity to the Union and helped it to sustain membership. Membership in 2001 numbered 300,000 families (table 9.1). The proportion of those who are farm families is not known but is believed to be higher than for the Grange or Farm Bureau Federation to be discussed later.

## EQUITY ORGANIZATIONS

J. A. Everitt in Indiana founded the American Society of Equity in 1902. Membership reached a maximum of about 40,000 in 1912. It was not a secret society with fraternal features, nor was it much interested in politics. It objectives were economic.

Everitt was publisher of a newspaper and author of *The Third Power.* The theme of this book was that farmers should organize in defense of their economic interests just as business and labor had done. The Equity, with its main strength in the upper Midwest, sought early to raise prices by holding farm products off the market.

The first holding action was attempted in 1903 on wheat and was followed by attempted price fixing for other crops. Results were not satisfactory, and many members suggested a reduction in acreage. A large amount of wheat was withheld from markets in the fall of 1906 with some success, but Everitt's vision of a successful holding action supported by several hundred thousand members never materialized. A faction under Everitt later formed the Farmers Society of Equity. It had no success with its price-fixing obsession and ceased to function in 1916.

Following the split with Everitt, the major group, under M. W. Tubbs, established its headquarters in Chicago. Its membership eventually was concentrated in Wisconsin, Minnesota, and the Dakotas. It emphasized cooperative marketing rather than price-fixing. The American Society of Equity had lost much of its dynamism through internal dissent, however. The Farmers Union and Nonpartisan League absorbed many of its activities and much of its membership over time. It was not a significant organization after 1917.

Its efforts to merge with the Farmers Union failed in 1910. Out of the attempted merger grew a new group, the Farmers Equity Union, organized by C. O. Drayton, a cooperative business organization with no fraternal features. It had an estimated 65,000 members in 1923. Sound business practices, with local cooperatives operating in conjunction with central facilities in a nationwide marketing plan, worked well for wheat Equity organizations that operated for decades in the Great Plains.

Organization of local cooperatives was a major accomplishment of the American Society of Equity and the Farmers Equity Union. In addition to grain marketing and livestock shipping, the American Society of Equity sold insurance and operated elevators, warehouses, meatpacking plants, flour mills, creameries, and retail stores. The Equity movement in the first two decades of the 1900s was unique. Unlike previous movements, it grew despite farm prosperity. The holding actions and price-fixing schemes were failures, but the business ventures were often financially sound and commanded member loyalty. Because the business ventures were most successful in times of prosperity, it follows that the organization would prosper in these times.

**The Equity Cooperative Exchange**

Another significant milestone in the farmers' movement was the Equity Cooperative Exchange. It started in 1907 in response to numerous grievances of farmers against the wheat trade. The intent was for the cooperative to own elevator facilities at the local and terminal level and to gain a seat on the Minneapolis Grain Exchange. It failed in the latter but did control marketing of much wheat, about eight thousand carloads in 1915. Spurred on by George Loftus, a fiery and informed business manager, it succeeded in securing many reforms in the grain trade, but the organization grew too fast and could not efficiently finance its operations. The Farmers Union absorbed it in 1934 (Taylor 1953, p. 409).

**The Night Riders**

Because Everitt's marketing plan appealed to tobacco growers, Kentucky was the strongest state in the American Society of Equity in 1905. The organization, along with tobacco-growers' associations, attempted to corner enough of the tobacco crop to control supply and to bargain collectively with buyers for a better price. The buyers foiled these attempts by paying a higher price for tobacco to "Hill Billies," who operated independently of the control organization. A small group of growers called the Night Riders operated like goon squads to force the buyers to bargain with the control group and to force the Hill Billies into the cartel. The Night Riders, whose activities were condemned by the Society of Equity, burned barns, destroyed tobacco plantings of growers not participating in holding actions, and destroyed properties of the tobacco companies. There were also cases of beatings and other intimidation.

## THE NONPARTISAN LEAGUE

Authur Townley, a former Socialist Party organizer, started the Nonpartisan League in 1915 in North Dakota. The grievances of wheat farmers were many, including a deep distrust of grading, storage, transport, and speculation in the grain trade. It was said of wheat that the farmer would sell a bushel and get paid for a peck, and the consumer would receive a peck and pay for a bushel. The middleman and banker were blamed for farmers' economic ills and were the common enemy. Encouraged by receptive farmers and enticed by high fees, organizers moved quickly, and the Nonpartisan League swept North Dakota like a prairie fire. It was said to have 200,000 members in 1918, mostly in North Dakota but also in surrounding states. Political power grew, and in 1918 the League elected its man for governor and other executive offices and gained control of both houses of the North Dakota legislature.

Among other accomplishments, the League established the Bank of North Dakota to provide low-cost loans, and a state-owned North Dakota Mill and Elevator Association to engage in processing and marketing of farm products and to establish warehouses and elevators. The legislative program of the Nonpartisan League established the necessary framework for state-operated home building and insurance.

Membership had begun to decline by 1921 for several reasons. Lack of adequate management and financial support in the face of formidable opposition from private enterprise and other groups caused the League's business operations to lose money. The leaders of the League became so involved in politics that the business organization and membership campaigns were neglected. Also, World War I brought charges of disloyalty and pacifism against the League and interfered with recruitment of new members. The Nonpartisan League (NPL) continued to be a political force in North Dakota. Candidates for public office continued to run on the Democrat-NPL ticket in the twenty-first century. The state-owned bank and the state-owned mill and elevator continued to function in the twenty-first century. These operated without NPL affiliation, however.

## AMERICAN FARM BUREAU FEDERATION

At the close of World War I, farm organizations were not very active. The Grange had a membership of approximately 550,000 persons, but the fraternal, secret approach was losing its ability to attract new farm members. The National Farmers Union had comparatively little influence; its membership was some 140,000. The American Society of Equity had 40,000 members, mainly in Wisconsin, and Nonpartisan League influence was waning as its membership fell from 200,000 in 1918 to 150,000 in 1922. Two decades of prosperity and the identification of the old organizations with past economic problems, many of which were no longer serious, had drained vitality from traditional farm groups. The old populist issues of monopolies and banks had lost some of their allure. Yet new problems were emerging, and the time was ripe for a new farm organization.

A new organization got its start from an unexpected source–the Cooperative Federal-State Extension Service. Local farm bureaus made up of educated, progressive commercial farmers had been organized to assist the county agent in education and demonstration of improved farming practices. These local groups were concerned not only with more efficient farming but also with the broader problems faced by farmers. These broad problems could best be approached from the state and national level. At a national meeting of county agents in Chicago in 1918 it was resolved that state and national federations of farm bureaus were needed to serve farmers more effectively. State bureaus were organized. At a national convention of these in 1919, Midwest advocates of a national federation that would deal with economic and business functions won out over eastern and southern groups who favored a more limited, educational role for a national federation.

The convention passed several resolutions including a call for economy in government and an increase in the limit on Federal Land Bank Association loans. These and other resolutions were clear and early evidence that the group would be involved in national public policy issues dear to commercial farmers.

In 1920 the constitution drawn up at the 1919 meeting was ratified at a national convention in Chicago, and James Howard of Iowa was elected president. By 1921, 42 state federations were in existence and by 1922 the American Farm Bureau Federation had a paying membership of 315,000 families. From its inception, the Farm Bureau, as it was called, represented commercial, middle-class farmers. Unlike previous groups, it did not cultivate the populist image of an oppressed, underprivileged class. Gray Silver was sent to Washington as a lobbyist and soon made the presence of the Farm Bureau felt in efforts to repeal daylight saving time, to regulate packers and stockyard companies, and to use government loans to increase farm exports.

The Farm Bureau was not initially committed to the McNary-Haugen Bills, but with a new Bureau president and new Washington lobbyist, the organization supported the third McNary-Haugen Bill in 1926. That bill sought a high price on farm commodities sold in the domestic market with remaining production to be dumped on the export market for whatever the market would bring. The third bill did not pass Congress. The fourth and fifth bills passed Congress but were vetoed by President Calvin Coolidge. None of the McNary-Haugen proposals became law.

Legislative proposals urged by the Bureau in 1932 included: (1) government guarantee of new deposits in all banks, (2) monetary reform, (3) restoration of price parity for agricultural products, and (4) relief for heavily mortgaged farmers in distress. The Bureau supported much of the New Deal legislation in the 1930s. The Farm Bureau claimed pride of paternity in creating the Agricultural Adjustment Administration (the federal agency administering farm commodity programs) and reaffirmed its faith in that organization in several national convention resolutions during the 1930s. The Bureau was especially active in securing farm credit legislation and was one of the first farm groups to recognize the folly of high tariffs and the need for reciprocal trade agreements to foster trade.

**Political Ties**

The Grange, the Farmers Union, and the Farm Bureau were often united in support of the McNary-Haugen Bills and the New Deal legislation. Farmers could present a united front in times of economic distress. The return of prosperity after 1940 eventually led the organizations down quite different roads, however. The Farm Bureau strongly opposed supply management, acreage diversion, and other programs of the Kennedy and Johnson administrations.

The Farmers Union and the Grange felt that the Farm Bureau had the undue advantage of support by the Extension Service and land grant colleges. (The ties between the Extension Service and the Farm Bureau were not completely severed until after World War II.) Despite some animosity by competing farm organizations toward the Bureau, the policies recommended by the major farm groups were not unduly different–all recommended legislative measures prior to World War II that were not out of line with traditional thinking. The Farm Bureau maintained a close liaison with business and commercial interests, while the Farmers Union worked more closely with labor unions. The Bureau developed a strong program of education, insurance, and cooperative marketing and by 2001 had a membership of 5.1 million families (table 9.1). An undisclosed but small percentage of these was farm families.

Changes in the membership of the Farm Bureau in recent years have profoundly changed the politics of the organization, returning it to its more populist 1930s position. Southern states have been especially successful in peddling insurance and associated membership to nonfarmers. A consequence is that in 2001 seven of the top nine states in membership were in the South. Tennessee, the top state, had 530,221 members, and relatively few of them were farm families. Voting is proportional to total membership, but those who vote are mainly commercial farmers. A consequence is the shift of power to the more populist, government-support-oriented South (see Guither, Jones, and Spitze 1984, p. 15). A manifestation of the shift was replacement of Dean Kleckner of Iowa with Bobby Stallman of Texas as president in January 2000–putting an end to a long line of conservative Midwestern presidents of the Federation. The Farm Bureau was a strong supporter of sizable transfers to agriculture in the 2002 farm bill.

**The Farm Bloc**

The index of farm prices fell from 211 in 1920 to 124 in 1921 (1910-1914 = 100) and was never over 156 during the remainder of the 1920s. This acute economic distress of farmers led to a resurgence of farm organization activity. Representatives of all the major farm organizations met with leaders of Congress and the President's Cabinet in 1921. The result was formation of a powerful bipartisan coalition of senators (and later representatives) that became known as the Farm Bloc. The Farm Bloc is not treated extensively in this chapter because it was not a farm organization but a bipartisan coalition of people in Congress concerned with farm interests through the national political process.

The outlines of legislation, worked out mainly between the Farm Bureau through its representative, Gray Silver, and the Farm Bloc, soon emerged.

Subsequently, laws were enacted in the interest of farmers dealing with packers and stockyards, futures trading, agricultural credit, cooperative marketing, and lower freight rates. With these high-priority measures of the Farm Bloc enacted, and with the economic crises of the early 1920s, the Farm Bloc began to lose some of its effectiveness by 1923. But the Bloc continued to exist at least into the 1940s. One could argue that the Farm Bloc exists in 2002. It is manifest in farm policy that is at once parochial and nonpartisan.

## THE COOPERATIVE MOVEMENT

The cooperative movement began as early as 1810 when dairy producers in Connecticut attempted to churn and market butter cooperatively (Taylor 1953, p. 472). Successful dairy cooperatives were established in Wisconsin in 1841 and in Oneida County, New York, in 1851. Cooperative marketing has since been extended to nearly all farm commodities.

Approximately one-third of farm output and a like proportion of farm input were marketed through cooperatives in year 2002. Membership in 1964 was 3.6 million (some farmers were members of more than one cooperative).

Establishment of cooperatives was an important activity of all the major farm organizations. The Grange, the Farmers Alliance, the American Society of Equity, the National Farmers Union, and the Farm Bureau organized cooperatives for two purposes: (1) to create a more efficient and orderly marketing system that would reduce middleman margins and (2) to control supply and regulate marketing so as to secure a higher retail price. In the former, the record has been mixed but overall must be called a substantial success. In the latter objective of supply control, farm organizations have been unsuccessful. That lack of success has been good fortune indeed for food consumers.

### The Sapiro Movement

One of the most successful cooperatives began as the Pachappa Orange Growers Association in California in 1888. Composed originally of eleven growers who pooled their fruit to sell to packers, the Association expanded in size and functions until in 1893 it was supervising the marketing operation up to the wholesale level. This organization led eventually to the California Fruit Growers Exchange (Sunkist) in 1905 and continued to operate under that name. The pattern of organization was repeated for other West Coast fruit cooperatives and eventually became a blueprint for the Sapiro movement. The pattern is as follows: Producers of a commodity enter into an agreement to pool their product for marketing. They incorporate and elect a board of directors that sets up a business organization and hires the management. The management in turn operates the marketing functions of receiving, processing, shipping, financing, storage, advertising, and selling. Members are required to sell only to the organization, and net returns are prorated back to each member according to the products that member supplied.

The successful marketing operations of cooperatives on the West Coast led the American Farm Bureau Federation to call a meeting of interested individuals in 1920 to hear Aaron Sapiro, a lawyer for several California cooperatives. He told the four hundred delegates, mostly from midwestern and eastern cooperatives, that the techniques used to market specialty crops in California could be applied to the major crops. Sapiro advocated a strongly centralized producer cooperative that would issue contracts to members for delivery of a specific volume of commodities, with penalties for noncompliance. Activities of the cooperative were to reach beyond first processing and storage and were to extend to warehousing and terminal markets. Prices were to be set and administered by the board of directors. Commodities that would not bring this price would be "dumped," presumably on the foreign market. He insisted that the monopoly cooperative should gain control of 90 percent of the entire crop to enable it to administer prices.

Accordingly, attempts were made to organize grain growers, livestock producers, cotton growers, tobacco growers, and others, but the groups were organized too hastily and with inadequate management. Farmers were yet too numerous, too diversified, and too independent for these strong measures. One after another, the groups collapsed, and the Sapiro movement for major farm commodities had failed by 1924 (Benedict 1953, p. 198). The philosophy lingered, however, and the basic system has continued for some specialty crops.

**The Capper-Volstead Act**

The Sapiro movement made it important to clarify the status of cooperatives in the light of the Sherman and Clayton Antitrust Acts. The Capper-Volstead Act, the Magna Charta of cooperatives, became law in 1922. It established the conditions under which an organization might be defined as a cooperative and to a considerable extent exempted cooperatives from the antitrust provisions of earlier Sherman and Clayton acts. Jurisdiction of matters that dealt with price-controlling activities of cooperatives was placed with the secretary of agriculture rather than with the Federal Trade Commission or the U.S. Department of Justice, the secretary looking more favorably upon cooperatives.

**The Federal Farm Board**

The Republicans answered the 1920s farm problem with the Agricultural Marketing Act of 1929. Its principal attack on the farm problem was to take the form of giant cooperative institutions. The Federal Farm Board, created under the act, was composed of eight members who were to authorize and support new cooperative marketing associations. Funds of $500 million were appropriated by Congress to be used by cooperatives to buy surpluses and support farm prices. The Farm Board, because of inadequate financing, was overwhelmed by the immensity of the farm problem. Cooperatives serve a useful function of efficient marketing and are loyally supported by members, but they were and remain today ill equipped to perform the vast chore of aligning supply with demand at an acceptable, stable price level for agriculture.

**National Council of Farmer Cooperatives (The Co-op Council)**

The National Council of Cooperative Marketing Associations was liquidated in 1926 and reorganized as the National Cooperative Council in 1929. The name was again changed in 1939 to the National Council of Farmer Cooperatives. The organization–from 1920 to the present–will be referred to here as simply the Co-op Council.

The Co-op Council (like the Grange, Farmers Union, and Farm Bureau) was one of the big four farm groups represented in the Farm Bloc. In 1955 it was composed of farm supply and farm-marketing cooperatives that included 116 separate affiliates, representing 5,000 local cooperatives serving a membership of nearly three million (McCune 1956, p. 54). It is a large organized farm group, although it is not ordinarily considered a farm organization as such. While cooperatives have an effective lobby and keep their members well informed, the interests of the many component groups are so diverse that the Council rarely takes a stand on national political issues and partisan politics. The Council has been called a "sleeping giant."

The National Council opposed the McNary-Haugen Bills of the 1920s and the commodity programs and acreage control schemes of the New Deal. The thinking that cooperatives could better deal with the problems that the commodity programs were attempting to solve motivated the Council's position. The Co-op Council has been a conservative voice in farm policy, closer to the Farm Bureau position than to that of any of the other major farm organizations. With Ezra Taft Benson as head during World War II, the organization found itself siding with the Farm Bureau, the Grange, and the U.S. Chamber of Commerce to fight Roosevelt's price ceiling policies and his measures to improve the housing and wage standards for migrant workers. Cooperatives had become large employers and were taking the business-management point of view.

## FARMERS HOLIDAY MOVEMENT

> We'll eat our wheat and ham and eggs
> And let them eat their gold. –*Iowa Union Farmer*

Farmers consoled themselves through the 1920s by saying, "It could be worse." And sure enough it got worse. The index of prices received by farmers, 211 (1910-1914 = 100) in 1920 and 148 in 1929, fell to 65 in 1932. The slump was no ordinary business recession but a disastrous depression. Per capita income of farm people averaged about one-third that of nonfarm people. The number of banks dropped from 30,000 in the early 1920s to less than 15,000 in 1933. In 1931 alone, 2,300 closed and wiped out savings for a large number of depositors. Only the farm economy was depressed in the 1920s; in the 1930s both the farm and nonfarm economies appeared hopelessly bogged down in a depression. Some farmers burned corn for fuel. Receipts from shipment of livestock to Chicago did not always pay the transport cost. Nearly a million farmers lost title to their farms from 1930 to 1934. The desperate situation led to protest in even the most conservative sections of the country.

The Iowa Farmers Union was the organized core that led to the "Farmers Holiday" rebellion. Union membership was only 9,600 in Iowa in 1932, a far smaller organization than the state's Farm Bureau, but its members were militant. The movement centered in Iowa but was active also in Nebraska, Minnesota, North Dakota, and other states. Milo Reno, even when not president, dominated the Iowa Farmers Union almost from 1918 when he joined it until his death in 1936. He was the most significant individual in the organization of the Holiday movement. Reno was a charismatic leader with an evangelical style and a strong promoter of a cause that preoccupied major elements of the Farmers Union–cost-of-production pricing. In 1927 Reno introduced a resolution to a Cornbelt committee, stating, "If we cannot obtain justice by legislation, the time will have arrived when no other course remains than organized refusal to deliver the products of the farm at less than production costs" (Shover 1965, p. 27). It took five years to realize the mandate of this statement.

He made occasional references to his plan in the *Iowa Union Farmer*, the Iowa Farmers Union newspaper edited by H. R. Gross, who much later (in the 1960s) was to be a very conservative Republican representative to Congress from Iowa's third district. Reno and John Bosch of Minnesota presented a resolution to the National Farmers Union convention in 1931 calling for a farm strike. The motion was soundly defeated. The holding action thereupon proceeded as an independent movement. But, "in a real sense, the Farmers Holiday Association was a strong-arm auxiliary of the Farmers Union" (Shover 1965, p. 35).

In February 1932, 1,500 farmers in Boone County, Iowa, met and pledged to "stay at home–buy nothing–sell nothing." Factions within the Farmers Union continued efforts to organize a strike and in May 2,000 farmers assembled in Des Moines, Iowa, to launch the Farmers Holiday Association as a national movement. Reno was named president. Organizers traveled through farm areas and secured pledges from a half-million farmers to support the forthcoming holding action. Plans to begin the strike July 4 fizzled; the strike officially began August 10, 1932. Sporadic action occurred in other places, but the focal point was Sioux City, Iowa. On August 11, as many as 1,500 farmer pickets stalled the movement of milk into the city. The strike was settled on August 20 with somewhat minor concessions from buyers. Picketing continued around Sioux City, however, and the movement became increasingly more violent as it spread to other locations. Pitched battles erupted between deputies and pickets at the Omaha city limits. A large number of arrests were made. A holding action by the Minnesota Holiday Association began September 21 and ended October 22. At least one picket was shot. The Minnesota action, like that in other areas, achieved very limited direct success.

Mass protests of farmers against foreclosure sales brought current economic issues into sharp focus. More farmers were involved in antiforeclosure activity than in any other form of protest in the 1930s. Crowds of from one hundred to two thousand insurgent farmers gathered to block foreclosure sales. These "spontaneous" gatherings, not led by an organized group, dotted the farmbelt. Some were so-called penny auctions, where the mob forced the property to be sold for a nominal amount so that it could be returned to the former owner (Shover 1965, pp. 77-81).

The exact impact of the farmer insurgence in the 1930s is difficult to judge, but the proven ability of farmers to use strong measures in pursuit of pressing needs undoubtedly influenced politicians and was a factor in subsequent legislation to relieve the acute economic distress brought on by depression.

## NATIONAL FARMERS ORGANIZATION

The farm economic picture was less than dark in 1955. The parity ratio (the ratio of prices received to prices paid by farmers) was 58 (1910-1914 = 100) in 1932 when the Holiday movement occurred; it was 84 in 1955. Nonetheless, prices and incomes were low in a relative, more recent perspective. As recently as 1951 the parity ratio had been 107. Net farm income fell from $14.8 billion in 1951 to $11.2 billion in 1955. But, most important, the price of hogs had dropped from $23.11 a hundredweight in 1953 to $13.10 a hundredweight in 1955. And hogs were very important to farmers in the cornbelt where the new movement was to be centered. Thus, it was hogs and relative farm recession rather than depression that moved farmers to protest. Drought in southern Iowa and debts accumulated since World War II also played a part.

The National Farmers Organization "started from wonderings aloud between a farmer and feed salesman in an Iowa farmyard one day in the late summer of 1955" (Brandsberg 1964, p. 4). The feed salesman, Jay Loghry, was the initial leader of a subsequent meeting of 35 farmers at Carl in southwest Iowa. A week later a meeting was held at Corning, Iowa, and 1,200 attended. The initial aim was to "unionize" farmers, but the term "union" received a very negative response, so the group took the name of National Farmers Organization (NFO) in September 1955. The NFO's only employee was Loghry, and its unpaid adviser was Dan Turner, former Iowa governor, who had called out the state militia to quell violence during the Farmers Holiday Movement. Turner, on the basis of his 40 years of experience, advised the NFO that it would not achieve its economic goals through strikes and violence but only through action by the U.S. Congress. Though Turner was a charismatic leader and attracted a large membership, the organization later acted entirely out of keeping with his advice.

The first national convention of the National Farmers Organization met at Corning, Iowa (national headquarters), in December 1955 with 750 in attendance. Oren Lee Staley was elected president. The organization listed 55,659 members after only three months of existence. In April 1956, the organization reported 140,000 members (Brandsberg 1964, p. 70).

The NFO, like the Farmers Holiday Association before it, first sought help in Washington to alleviate economic problems. When visits to Washington and attempts to elect members to Congress failed, the 1957 NFO convention looked to collective bargaining as a more persuasive solution.

Trial "testing" actions–small-scale holding actions of limited duration–were begun in 1959. Another was held in 1960. A final test holding action was held in 1961 with encouraging results.

Some 13,000 members attended a 1961 meeting, and 20,000 attended a 1962 meeting in Des Moines in preparation for the all-out livestock holding action planned for 1962. The action was slated to begin September 1, 1962. Possibly in anticipation of the action, farmers flooded the market, and on the Friday before Labor Day,

Midwestern stockyards received the heaviest flow of cattle in 14 years. Prices, however, did not fall, and the price for prime cattle in Chicago reached $32.50 a hundredweight, the highest it had been since April 1962 (Schlebecker 1965, p. 208). Processors were trying to stock up on slaughter cattle before the strike.

Initial success of the strike that began after the Labor Day weekend was startling. By September 6, 1962, interior livestock market receipts had fallen 50 to 75 percent. Livestock market prices rose. Approximately six hundred packinghouse workers were laid off by September 7 in Iowa alone (Schlebecker 1965, p. 208).

Reports of violence soon appeared. Farmers said rifle shots were fired at trucks hauling livestock to market. Tires were slashed. Nails were spread on highways.

NFO membership is secret, but one estimate placed membership at 180,000 or more farm families in the early 1960s (Tontz 1964, p. 146), too few to impact markets, but the holding in 1962 was supported by a great many nonmembers. Furthermore, it was commonly believed that NFO members had smaller than average farm units. One investigation found that on the average NFO members were better educated and had larger farms than Farm Bureau members (cf. Schlebecker 1965, p. 212). The membership fee of $25 discouraged very small farmers from joining.

Enthusiasm began to wane in the second week of holding. Nonmembers were the first to sell. In terminal markets for the week ending September 15, cattle and calf receipts were 235, 464 compared with 176,048 the week before and 231,658 for a like week a year earlier (Brandsberg 1964, p. 115). Hog receipts numbered 155,000 during the first week of the strike and 265,000 the second week at Iowa and southern Minnesota markets.

Meanwhile, NFO members continued to staff their checking stations at crossroads and markets. Considerable peaceful picketing was carried on in stockyards. Oren Staley continued to use his persuasive oratorical powers to maintain member morale. But some members began to sell their livestock. In the fourth week, leaders considered having dairy farmers enter the holding action. The momentum of the strike had been lost, however, and on October 2 the drive was "recessed," with a promise to resume it at a moment's notice. It was not resumed. The action had failed in its goal to sign contracts with enough processors to assure farmers of a market at adequate prices. Nevertheless Staley stated that it was a "breakthrough, the starting to sign contracts with processors. . .because this is the first time that farmers had contracts signed that would determine future price levels" (Brandsberg 1964, p. 132). Little is known of what contracts were signed with meat packers. The effect on prices and incomes was insignificant.

The NFO in 1963 turned its attention to collective dairy bargaining. Contracts were signed with some creameries to deliver milk. This was a considerably easier operation than with livestock because the market was localized. The processors were cooperatives with some NFO members and sympathizers on the board of directors, and the numerous dairy processors and producers were already accustomed to operating under federal milk marketing orders. At Annandale, Minnesota, after a creamery had signed a contract with NFO, the dairy processor to whom the creamery delivered its skim milk would not accept any more skim milk unless the creamery legally terminated its contract with NFO. The result was a gathering of 2,000 farmers at which some 40,000 gallons of milk were dumped in the ditch in March 1963. Subsequent minor holding actions in soybeans, grains, and livestock came and went with little impact on prices or the economy.

The National Farmers Organization was nonpartisan in its politics from the beginning. Nevertheless, it had been formed partially as protest against the policies of the Republican secretary of agriculture Ezra Benson. The NFO received Benson's successor, Orville Freeman, with much fanfare at one of its meetings and was clearly closer to the Democratic Party than the Republican Party in its philosophy. The term "labor union" was anathema to many Midwest farmers, but the NFO received early inspiration and even some financial support from labor unions. The Farmers Union was clearly in sympathy with the NFO, and an attempt was made to merge the two organizations in Iowa.

Opponents of the NFO included the processors and food chains, the Farm Bureau, the National Livestock Feeders Association, and the American National Cattlemen's Association (cf. Brandsberg 1964, pp. 224-36). This created some ambivalence among farmers, since numerous farmers were members of both the Farm Bureau and NFO. It cannot be said that the groups mentioned above caused the failure of the holding actions. It was rather farmer independence and the resource and market structure of agriculture that wrecked bargaining attempts. Perhaps the strong measures taken by Secretary Freeman and the Congress to control production and maintain prices were a telling blow, because these measures redirected the attention of the farmers to Washington for alleviating the farm problem in the early 1960s.

The NFO in March and April of 1967, with a reported quarter of a million membership, showed renewed vigor in a 25-state holding action to raise the price received by farmers for milk from 10 to 12 cents per quart. Although thousands of gallons of milk were poured on the ground and the farmers materially reduced the flow of milk to markets, the holding action had less effect on the market than anticipated. Diverting milk formerly used to produce cheese and butter to the fluid milk market averted a serious shortage of milk.

The movement seemed destined for complete failure, but in April the strike was given new impetus when two labor unions, the meat cutters and teamsters, agreed to honor the NFO picket lines in Nashville, Tennessee. The action was so effective that the milk supply was cut off for all but emergency uses. It is interesting to note that a union group that the NFO had in its earlier days so thoroughly rejected made the turn of events possible. The holding action, even before it was joined by labor unions, was successful in obtaining contracts that satisfied price demands from processors in some localities. Nevertheless, the NFO had not come to grips with the long-term problem of how to avoid excess supply in a market stimulated by prices above the equilibrium level. As a means to survive, the NFO shifted to a strategy of providing conventional marketing services to farmers. That strategy was only partially successful, however, and the organization had nearly withered away by the twenty-first century. A few National Farmers Organization dairy cooperatives continue to operate using conventional marketing methods.

**AMERICAN AGRICULTURE MOVEMENT**

With the parity ratio 64 (1910-1914 = 100) in September 1977, unrest again emerged among farmers. Falling grain prices added to the economic distress caused by cattle prices that had been depressed for two years. Many farmers, especially the young, had insufficient cash flow to service loans on land and machinery purchased in the boom of 1973-76. Input price inflation and drought in some parts of the country, notably in the High Plains and Southeast, made matters worse.

In September 1977, a truck driver, overhearing a conversation among disgruntled local farmers in a Baca County, Colorado, coffee shop, from his labor union experience urged the farmers to "do something about it." The farmers drew up and distributed handbills advertising the fledgling movement. Shortly, thereafter, a large number of farmers formed a tractor caravan to Pueblo, Colorado, to meet with Secretary of Agriculture Robert Bergland. They received little satisfaction.

Their immediate concern was the price of wheat. One of their leaders, Bud Bittner, stated: "The only problem in agriculture is price; we all deserve a wage." Farmers promised a strike by December 14 if demands were not met for 100 percent of parity for all farm products to cover the "real cost" of producing crops. A pamphlet stated that if demands are not met "we will not sell any agricultural products. We will not produce any more agricultural products. We will not buy any agricultural equipment or supplies." (see Box 9.1)

Widespread dissatisfaction with farm economic conditions provided fertile ground for protest, and the strike movement soon blanketed the High Plains and then the nation. Parades of tractors, called tractorcades, displaying American Agriculture Movement banners, became commonplace in small towns and large cities including Washington, D.C. Rallies of farmers were widespread, some drawing as many as 30,000 people.

To emphasize its grassroots origins and to attract members of traditional farm organizations, the movement avoided electing officers, paying salaries, and becoming a structured organization. By early 1978, the movement reached a crescendo in pressing its fight for 100 percent of parity. Thousands of farmers became involved in intimidating politicians and local merchants, occupying a Rio Grande bridge to stop fruit and vegetable imports from Mexico (many farmers were arrested), demonstrating in Washington, D.C., breaking down the door into the U.S. Department of Agriculture administration building in Washington, and finally filling the hearing rooms of Congress. The media cooperated by elevating local incidents into national news. Never before had so many farmers become so militant so fast.

The movement was another episode in populism and its goals originated out of frustration rather than careful reasoning. Its crusade was for 100 percent of 1910-14 parity, an obsolete concept. Its proposal that every buyer be required by law to pay 100 percent of parity for any farm product would have generated massive food production overcapacity, turned agriculture into a police state to enforce provisions of the law, tripled land prices, and sharply raised consumer food prices. (A national opinion poll showed that the majority of consumers was sympathetic to raising farm prices but only to the extent of raising food prices no more than 5

**Box 9.1**

Some of the fondest memories of my professional career arise out of my numerous interactions with farm populists. On a personal level, I have found them to be congenial individuals. In groups, mob psychology sometimes brings out the worst in them, however.

I spoke at several conferences either held for or dominated by American Agricultural Movement (AAM) activists in the 1970s and 1980s. Activists were upset with the lessons that I related to them from historical accounts of protest movements. I observed that farm protest movements often began with attempts at food supply management to raise farm prices and incomes. These attempts failed. Protest movements then faded away unless the movement was institutionalized into an organization that provided insurance, cooperative supplies and marketing, and other services to meet member needs. One activist upset with that advice (which continues to hold today) remarked to me after I spoke that "We [The American Agricultural Movement] will gain power and, when we do, we will remember our friends." I was not one of their "friends." The last thing they wanted to hear from me was "my best professional judgment," even if it was correct.

Although I was a publicly employed economist paid by taxpayers from the 1950s to 1988 and held an endowed chair from 1988 to 2000, populist farmers felt that my job was to serve farm interests even if farmers gained less than consumers and taxpayers lost. I viewed my job as serving the public, however. I looked for that happy circumstance where serving farm, consumer, and taxpayer interests did not conflict, as in public support of basic agricultural research.

In late 1978, a National Farm Policy Summit was held in College Station, Texas. The sponsoring Agriculture Council of America sought to keep the meeting peaceful at a time of American Agriculture Movement protests by imposing a registration fee of $90 to keep demonstrators out. The ploy did not work. Some 200-300 AAM activists stormed in without paying, and Bud Bittner, their spokesperson, immediately grabbed the microphone, declaring AAM was in charge. I was especially nervous because the AAM activists, mostly from the Texas Panhandle, had obtained my presentation paper from Oklahoma AAM members with which I had been seminaring. The Texans didn't appreciate my paper, which contended that farmers were experiencing a cash-flow problem caused by inflation (they were paying 12 percent mortgage interest while earning 4 percent on land; the difference made up by an 8 percent or higher land capital gain that unfortunately could not be realized until that land was sold). The problem was not a low rate of return that could be solved by 100 percent of parity because higher commodity prices would raise land prices, intensifying cash-flow problems for new farm owner-operators and renters.

We worked out a deal. AAM's "theorist," a veterinarian named Dr. Schroeder, would be given equal time after I spoke. Professor Theodore Schultz, a Nobel Prize-winning economist at the University of Chicago, chaired the session. To avoid confrontation, he did not allow any exchange between me and AAM advocates in the discussion following formal presentations.

My relationship with AAM appeared to be adversarial, but national AAM leaders asked for and received a meeting with me at Stillwater, Oklahoma, sometime after the National Farm Policy Summit. During that polite meeting with AAM leaders, I discovered they had no interest in "structure" issues (policies affecting farm size, numbers, tenure, and agribusiness concentration) prominent at that time with Bob Bergland as secretary of agriculture. AAM advocates were commercial farmers, not small farmers, and they did not want government policies to stand in the way of profit or expansion–they just wanted 100 percent of price parity.

percent.) Actually realizing 100 percent of parity would have generated huge windfall gains for wealthy landowners at the expense of less-wealthy consumers. "Full" parity would have hastened the demise of the traditional family farm because only a few wealthy landowners and corporations could finance the capital required to own or operate an economic farming unit.

Movement followers heavily lobbied a bill introduced in Congress by Senator Robert Dole of Kansas to provide 100 percent of parity price for produce of farmers who diverted sufficient acreage. The bill narrowly passed the Senate but was soundly defeated in the House in April 1978. Hundreds of farmers lobbying in Washington packed their bags in disillusionment and went home to plant crops. Improving farm prices also encouraged farmers once again to "turn in their swords for plowshares."

The strike, like previous holding actions, had virtually no impact on farm production and marketings. The formation of the organization was poorly timed. The Food and Agriculture Act enacted into law in September 1977 had exhausted farmers' political capital. Winter wheat had been planted by December and producers were reluctant to plow under their investment. Farmers most concerned about low prices could least afford to strike because they could not withstand the financial setback of no crop when burgeoning reserves dampened chances of raising wheat prices. The principal benefit to farmers was national news media coverage of the economic plight of farmers.

The coverage plus pressure for 100 percent of parity (later reduced to 90 percent of parity) placed on lawmakers by hundreds of farmers encamped in Washington brought some success. Secretary of Agriculture Bob Bergland expanded acreage diversion and made other changes in programs that he said would add $3-4 billion to net farm income. With the hope of farmers managing supply to obtain a better commodity price once again dashed, and with no other insurance or other business benefits to offer farmers, the AAM like so many populist efforts before watched farmers go back home to farm. Protesters returned to the fold of the Farm Bureau and Farmers Union whose membership they had never relinquished. Few remnants remain of AAM in the early twenty-first century.

## SUMMARY AND CONCLUSIONS

American history is littered with unsuccessful efforts of farmers to become price makers instead of price takers. American history also is replete with legislative measures to control production and improve economic conditions for farmers. This chapter evokes a raft of cliches: "There is nothing new under the sun; history repeats itself; those who do not learn from history are condemned to repeat its mistakes."

History teaches that farmers have not always conformed to the basically docile, democratic image of Currier and Ives paintings. The process of commercialization was unsettling. Neither farmers nor the public understood the exigencies of economic development characterized by technological change, economies of size, and Schumpeterian creative destruction. They reacted against real and imagined ills that a more enlightened educational and legislative process might have prevented.

A characteristic pattern of each protest movement is apparent: Low farm prices (not necessarily in absolute terms but relative to recent highs) and growing feelings among farmers that they were exploited by nonfarm groups helped create a new organization or commandeer an established one to deal with the issues. The movement was likely to be led by a nonfarmer–feed dealer, newspaperman, Department of Agriculture employee, socialist party organizer, or former grain exchange worker. Early attempts to control food supply and bargain collectively quickly failed but excited farmers and frightened consumers. Formation of cooperative businesses and other efforts to increase efficiency of farming were attempted, but impatience led to seeking legislation as the quickest form of redress for wrongs. When legislative attempts failed, the entire organization including its business interest was stifled and the demise of the organization was as complete, if not as dramatic, as its rise.

The protest movement has been confined mostly to the major cash crops and dairy farming and has been located in commercial farming areas, primarily the Midwest. Farmers formed marketing organizations to reduce the bite of market middlemen, they formed farm supply cooperatives to reduce costs of inputs, and they sought legislative redress for alleged failure of the economic system.

Despite the fact that farmers constituted the largest single industry in society, the nation was never in jeopardy from farm insurrection. The fact that only a small percentage of farmers at any one time was engaged in protest supports the conclusion that the goals and values of most farmers are closer to democratic capitalism than to farm fundamentalism and unbridled populism.

Protestors were not revolutionaries. They had a basic faith in the market mechanism, but they sensed a conspiracy against them. In their view it was the bureaucrat and monopolist who had distorted the market rather than the failure of the basic price system and free market structure that was to blame for economic ills.

With increasing economic education and experience, farmers are becoming more sophisticated in knowing the limits of supply management. Farmers have found it difficult to form effective organizations to control production for several reasons. First, the atomistic, widely dispersed structure of farming makes it difficult to coordinate activities. Coordinating farmers is like herding cats. Farm products are perishable and, unlike factory labor, either accumulate or deteriorate during a strike. The more effective a holding action is in raising prices, the greater the encouragement for farmers to not participate and to sell their products during the strike. Processors who enter into contracts to take only the products of contracting farmers at prices above the market face a reduction in profits and possible bankruptcy. Contracting processors usually must sell on the same market as firms securing raw materials at lower prices.

Organizations have learned certain lessons. One is that cooperatives are more successful in prosperous times than in depressed times. And legislation has more appeal in depressed times than in prosperous times. Thus, a wise organization will not dissipate its energies in a blaze of politics in good times nor will it seek to solve problems of farm distress with cooperative and other business activities in economically depressed times. A durable farm organization must have as its foundation solid, efficiently managed business enterprises to provide services the farmers demand, whatever the economic and political climate.

A commonly held view is that farmers' organizations are most successful when farmers experience distressed economic conditions. This seems to be true for the nineteenth century, but not since. While new organizations may be likely to form as protest movements, the established organizations depend for membership on the success of insurance and other business operations. These succeed best and farmers find membership dues easiest to pay in prosperous times.

Holding actions and other farm protest movements are tired solutions to farmers' economic ills. Economic ignorance has been costly to farm protestors. The ends including better living standards that desperate farmers sought have been achieved, but not with the populist ideology and actions pursued by protestors. The irony of ironies is that free enterprise markets, which protestors so despised, have brought high living standards to the farm and nonfarm economies alike.

Populism draws deeply from the myth that farm problems are caused by conspiracy of "middlemen" (agribusiness), by the Trilateral Commission, or by other bogeymen including government. In fact, there is no conspiracy. Low and unstable profit margins are expected in an industry characterized by low short- and intermediate-run income and price elasticities, rapid technological change, economies of size, and frequent and unpredictable shocks from nature and commodity and business cycles.

The surest defense against such perils facing farm operators is the ability to adapt to change (see chapter 10 and Tweeten 2002). The solution does not rest in self-help supply management or slower technological change. Rather, operators stay competitive through competent farm and risk management and marketing. Able operators of commercial farms have earned favorable returns in every decade since the 1930s and will do so in future decades (although not every year) with or without government commodity programs. Operators of farms that are undersized or inefficient will earn low returns but they can remain on the farm if they have outside means to pay for their hobby.

Unfortunately, not everyone who wants to farm can be successful in that occupation–whatever public policy is pursued. Even more unfortunately, society has not faced up to the fact that many would-be farm operators will fail. Society has not provided the safety net of counseling, job information and training, relocation assistance, and the like that would help people who are unsuccessful in farming to adjust to new opportunities in other lines of work. Providing such investments in human capital could diminish tendencies of farmers to seek solutions in failed populist ideology and protest.

**REFERENCES**

Benedict, M. R. *Farm Policies of the United States, 1790–1950.* New York: Twentieth Century Fund, 1953.

Brandsberg, G. *The Two Sides in NFO's Battle.* Ames: Iowa State University Press, 1964.

Davis, L. E., J. R. T. Hughes, and D. M. McDougall. *American Economic History.* Homewood, IL.: Irwin, 1965.

Guither, H., B. Jones, M. Martin, and R. Spitze. *U.S. Farmers' Views on Agricultural and Food Policy.* North Central Regional Extension Publication 227. Urbana: University of Illinois, 1984.

McCune, W. *Who's Behind Our Farm Policy?* New York: Praeger, 1956.

Schlebecker, J. T. "The Great Holding Action: The NFO in September, 1962." *Agricultural History* 39 (1965): 204-13.

Shover, J. L. *Cornbelt Rebellion.* Urbana: University of Illinois Press, 1965.

Taylor, C. C. *The Farmer's Movement.* New York: American Book, 1953.

Tontz, R. L. "Membership of General Farmers' Organizations, United States, 1874-1960." *Agricultural History* 38 (1964): 143-56.

Tweeten, L. "Farm Commodity Programs: Essential Safety Net or Corporate Welfare?" In *Agricultural Policy for the 21st Century*, edited by L. Tweeten and S. R. Thompson. Ames: Iowa State Press, 2002.

Tweeten, L. *Foundations of Farm Policy.* 2d ed. Lincoln: University of Nebraska Press, 1979.

# 10

# Summing Up: Costs and Cures

## INTRODUCTION

American agriculture is a bright and shining beacon, the envy of the world, an exemplary model of science, agribusiness, and farmers coming together to bring abundant food to people at home and abroad for richer and fuller lives. This volume, however, is about a darker side of agriculture that peddles snake oil, enmity, and violence.

The two groups, agricultural populists and radicals, considered in these pages could hardly seem to be more different. Populists are part of the agricultural establishment, although often on the fringes. Many are farmers or of a farm background. Radicals, on the other hand, are agriculturalists only in the sense that they are passionate about what they consider to be seminal issues of food and agriculture, and devote their time, energy, and intellects to their political agendas for change. They are outsiders to and, indeed, mostly disdain the agendas of the agricultural establishment.

So why are these strange bedfellows crammed together between the covers of this book? The foregoing chapters make clear that each group is adept at devising and promoting myth. And both groups use that mythology to generate animosity toward agribusiness and to further their political agendas. Both groups play on public fears. Populists claim the nation will run short of food and farmers because farm commodities don't earn a fair price. Radicals make similar apocryphal claims because people despoil the environment or have too many children.

The two groups differ sharply regarding violence. Populists have a long historic record of scattered violence, but they have learned that their political agenda is better served by other strategies. In contrast, radicals of nearly every kind attempt to further their agendas for agriculture with violence.

Unpredictable, unlawful acts of violence toward property and people to pursue political ends constitute terrorism. That terrorism sometimes is intended to intimidate the public and bring public policy favored by perpetrators; in other cases terrorism is retribution for alleged wrongs. Whatever the intent, the result often is costly in income, property, lives, and personal welfare.

In this final chapter, I review some of the costs to society of peddling myth, emotion, and violence regarding agriculture. The final section suggests policy initiatives to deal with the growing problem of agricultural terrorism originating from domestic and foreign sources.

## COUNTING THE COSTS

Not all costs imposed on society by radical and populist agriculture can be measured. The following accounting probably underestimates full costs by a large margin. My accounting does not include the billions of dollars spent by public and private entities countering misinformation and protecting against terrorism. Protective measures range from extra locks and insurance to hiring more security officers.

### Antiglobalists

Antiglobalists' policies, if implemented, would raise worldwide food costs and lower personal and national incomes by sharply curtailing international market flows of goods, investment, technology, and knowledge. The overall impact cannot be quantified, but almost certainly would condemn the 800 million food-insecure and the 1.2 billion poor people in developing countries to the status quo at best. As noted in chapter 3, failure to end current barriers to agricultural trade alone could cost the world over $50 billion per year or over $1 trillion in perpetuity in forgone national income.

Antiglobalists contend that their concern is economic equity and the environment. But analysis reported in chapter 3 confirms that benefits of economic growth in fact do "trickle down" to help the poor. Economic growth is not a luxury, it is essential to lift people out of poverty. Without a "pie" of income, issues of equitable division of benefits are moot. Studies also show that higher per capita income attainable with globalization reduces birthrates and population growth. Eventually, higher per capita income reduces total and per capita environmental degradation. Greater opportunities and rights for women and minorities attend higher per capita income.

So-called antiglobalists are in fact globalists when promoting their own policies for the world. "Antiglobalists" seek global policies to protect human rights, the rights of women and children, and the right to food. They want to preserve local culture–the ability of each society to "write its own stories." But the anti-economic-growth and antitrade policies they promote work against commendable efforts to better the lot of women, children, minorities, the poor, and the environment. Economic progress attainable only through globalization makes many of these worthy ends attainable. That is, richer countries provide rights that antiglobalists (ironically residing in rich countries) seek for poor countries.

**Radical Environmentalists**

Radical environmentalists predict doom for the world's inhabitants because of misuse of the environment (see chapter 4). Continuation of current policies, they say, will further immiserate a world that has too little food and natural resources, and too much pollution and congestion.

Reality is quite different as highlighted in chapter 4. Compared to previous generations, our generation lives longer. We also live better: we eat better, are healthier, have more leisure time, have more purchasing power, have more entertainment options, suffer less in the heat and cold, and face less drudgery on the job. Although poor nations lag, people there too are making progress and face ever-lower incidence of poverty and hunger. The *standard model* briefly outlined in chapter 3 offers food security and the means to end poverty for any nation willing to embrace it.

Progress is sustainable and is not coming at the expense of the environment. Technology and other forms of knowledge originating from human ingenuity and judicious investments in education and science have multiplied real output of goods and services available per unit of natural resources. Thus, living standards can continue to rise indefinitely per capita even as natural resources are drawn upon. American agriculture has made huge strides apparent in declining soil erosion, safer food and water, and productivity gains freeing environmentally sensitive cropland for wildlife biodiversity and for recreation.

According to estimates of the world supply-demand balance reported in chapter 3, future food abundance at falling real prices cannot be taken for granted. Also, global warming poses very real challenges. Implementing radical environmentalist's solutions would be counterproductive, however. Solutions can best come with input from science, education, human ingenuity, public discourse, and the market.

Many of the solutions proposed by radical environmentalist show disdain for economic benefit-cost analysis and the market, despite telling evidence that "cap and trade" emission markets efficiently reduce pollution. Free inquiry and science, by developing technologies that reduce costs of producing "clean" energy, contribute mightily to a better environment and higher living standards alike.

**Luddites**

Despite their best efforts, Luddites will not stop technological progress. If they could stop or roll back technology, the result would be calamitous as documented in chapter 5. In the United States, for example, productivity allowed farm food output to quadruple while aggregate farm production input remained at nearly a constant level through the twentieth century. That means that producing today's output with technology of a century ago would require as much as $1 trillion more farm input annually. The consequent high cost of food required to pay for more farm input would impose a special burden on low-income consumers.

To be sure, agricultural technology released much labor from farming. Development diminished the farm share of the nation's population, employment, and income. Luddites scorn that transition. But the rise of civilization is largely a history of increasing labor productivity and attendant release of labor from agriculture. People freed from agriculture supplied the education, health, entertainment, and other services essential to quality of life. Luddites would stop that process by forsaking economies of size from larger farms and agribusinesses and forsaking new bioengineering and information technologies.

### Animal Rightists

Animals are integral to Western culture. Animals have provided draft power, food, clothing, transportation, and companionship. Perhaps no radical agriculture group challenges Western culture more than the animal rightists. Scientists, including economists, are sometimes frustrated in addressing animal rights issues because they entail beliefs and values that people are entitled to hold. Animal rightists are not so inhibited, and promote their messianic vision for all of society by any and all means as observed in chapter 6.

Animal rightists advocate vegetarian diets partly because, compared to an omnivorous diet, plants can feed people with fewer resources. One response is the correct assertion that the earth has plenty of resources now and in the future to include animal products in the diet. Second, some lands are unsuited for crops and are best utilized as grazing for animals. Third, people enjoy eating meat when they can afford it. People can consume a healthy diet with or without animal products, but they freely choose to include more meat in their diets as their income rises.

Some animal rightists use violence to impose their preferences. The cost to farmers has been substantial. Terrorism is not going to achieve the vegetarian, fur-free society animal rightists seek. To be sure, after being intimidated by animal rights organizations some food firms have dictated that their animal product suppliers follow "green" animal production practices. Whatever its success, extortion by animal rightists imposes higher food costs on the majority by an unelected minority. Other, less confrontational means to bring change offer promise.

Reasonable people rightly will differ on standards of proper care and use of animals. Certain standards avoiding cruelty have wide appeal and rightly have become law. Beyond that, voluntary labeling, allowing consumers to buy range-fed or other preferred type of animal production practice, can achieve free choice in the market. Thus, values of the few need not be imposed on all. Voluntary labeling (with proper certification) allows a pluralistic society to flourish at affordable costs.

### Populists

Agricultural populists like radical groups considered in this study peddle myth and emotion (see chapters 7-9). However, as noted earlier, agricultural populists differ from the radical groups in several ways. Populists rarely resort to violence. Radicals ostensibly act in the interests of others; populists mostly just want a better price for their farm products.

Populist mythology has generated much enmity toward agribusiness. Politicians have listened, and the result in some cases has been regulation of agribusiness. Some of the more onerous regulations imposed on interstate commerce have been removed, but efforts are under way by Congress to introduce costly new constraints. Proposals, if enacted into law, would severely restrict production and marketing contracts and limit farm and agribusiness firm size, mergers, and acquisitions. The result could compromise efficient performance of agriculture and agribusiness. Measures would constrain ability of firms to reduce costs by adopting new technologies and achieving economies of size. If imposed, populist measures to control agribusiness will bring higher food costs to consumers but will not raise farm prices, income, wealth, or rates of return. These measures will not save family farms.

The success of tomorrow's farm economy rests neither with these populist solutions nor with more government payments, new product uses, exports, value-added enterprises, or improved technology–though many of these have merit. The economic success of farms depends on their ability to adjust to emerging economic conditions, whatever those conditions may be. In the future as in the past, with or without government commodity programs, adequate size commercial farms will experience favorable economic outcomes on average. They will not do well every year. Such conditions of instability will continue to plague farmers. Fortunately, the private sector offers attractive risk management tools such as forward contracts, insurance, and inventory management to help farmers cope.

Populist nostrums are especially expensive to society. Commodity programs alone cost taxpayers approximately $20 billion and the public at large $5-10 billion per year. Adding costs of agricultural trade barriers supported by antiglobalists and populists brings overall costs to much higher levels.

## AGRICULTURAL TERRORISM

Agriculture has experienced a large number of terrorist attacks, but has not experienced a terrorist incident causing anywhere near the damage inflicted on the World Trade Center towers on September 11, 2001. Although the direct cost (mostly in destroyed property) of terrorist attacks on agriculture is measured in millions rather than in billions of dollars, indirect costs in lost productivity and food security from forgone research has been substantial. Research offices, laboratories, and experimental plots have been frequent targets.

Law enforcement needs to take agricultural terrorism much more seriously, in part because the food supply is such an easy target for terrorists. The state and federal law enforcement response to agroterrorism has been lackadaisical. Few perpetrators have been apprehended. The nation knows comparatively little about the infrastructure and modalities of agricultural terrorists. The Federal Bureau of Investigation and Central Intelligence Agency must play a central role in thwarting attacks because terrorist cells operate across state and national borders. Greater effort is warranted for counterterrorism measures such as infiltrating cells, coordinating among agencies, and funding of training.

## A MODEST POLICY INITIATIVE: REINVIGORATING THE SOCIAL CONTRACT

Confronting radical and populist agriculture requires both carrot and stick. Those who cannot get along together regarding food and agriculture might well reconsider a fundamental construct: the *social contract*.

Social philosopher John Locke (1632-1704) proposed the concept of a social contract centuries ago. An informal, unwritten social contract always exists between people and government. Ideally, under the contract, individuals obey laws and accept and adjust to technological and economic change in exchange for the state's protection of property, person, and ballot. A bad contract is manifest in fatalism, distrust of institutions, and social problems that in the extreme bring revolution.

The Swiss-French philosopher Jean Jacques Rousseau (1717-1778) concluded that states ultimately survive because they fulfill the informal, unwritten social contract with the people they govern. The social contract between the government on the one hand and agricultural radicals and populists on the other is frayed. No competent government can long subscribe to a contract satisfying the objectives of populists and radicals, however. Because many of their demands are outrageous or contradictory, a contract with them would violate the terms of the contracts with other people. Nevertheless, some common ground can be found to strengthen the social contract for all.

Social philosopher Thomas Hobbes (1588-1679) is famous for his description of human life in a *laissez faire* natural state without government as "solitary, poor, nasty, brutish, and short." At the other extreme, the Hobbesian state describes state-run societies such as the Soviet Union where markets and individual freedom are suppressed. Thus, the social contract is of high quality only if the individual and the state perform their proper roles.

The social contract with agricultural populists and radicals can be improved despite the limitations. An example is macroeconomic policy. Populists have drawn much strength from the economic instability brought on by incompetent monetary and fiscal policies that, for example, contributed to and extended the Great Depression and the farm financial crisis of the early 1980s. Incompetent regulatory policies of the late twentieth century made possible the StarLink corn fiasco and other regulatory problems in attempts to protect the environment. So the first requirement is for government to operate as competently as possible under a representative, responsive democracy.

Of course, even competent governments cannot serve the needs of all people at all times. The inflexibility of the "all or nothing" policies of the government suggests advantages of giving greater opportunity for people to express their choices in the market. Every consumer need not be forced into a lockstep policy of eating only organic foods, only nongenetically engineered foods, or only domestically produced foods. As emphasized earlier, certification and voluntary labeling of foods can help the market to meet uniquely the needs of consumers with different cultural preferences and tastes. Letting consumers register their preferences in the marketplace allows plurality and individualism so prized by Americans to flourish.

Finally, the importance of education and research cannot be overemphasized. PAs and RAs recognize the importance of the media and have used them more effectively than has the conventional scientific establishment. That needs to change if the costs of radicalism and populism noted in this volume are to be averted.

Because policies advocated by radicals and populists do not serve even their own stated objectives in many cases, it would appear that an educational effort could be effective in exposing inconsistencies that would cause extremists to change their positions. Their thinking is not easily changed (to ease terms of the social contract), based as it is more on their unique ideology of beliefs and values than on analysis.

Thus, this book scrutinizing radical and agrarian populist philosophy, behavior, and positions is directed especially to the uncommitted. This volume is only a start. Education needs to reach into the home, school, and workplace. That educational effort will be most effective if it is buttressed by thorough research that, like the educational effort, is as objective as possible. Radical and populist agriculture ideas and beliefs need to be confronted whenever possible by dialogue, debate, and discourse.

## INTERNATIONAL AGROTERRORISM

International agroterrorism is defined as intentional acts of damage to rural people, property, or the food supply of this country by foreigners for political reasons. Fortunately, such acts have been few and hence received little attention in this volume. Canadians in the North American Animal Liberation Front are believed to have vandalized numerous fur farms in the United States in recent years. A Florida university professor informed the CIA that a citrus cancer outbreak was the result of a Cuban biological weapons program–a claim that could not be substantiated after investigation (Monterey Institute of International Studies 2002).

### Why Agroterrorism?

Agriculture and the food supply would be expected to be attractive targets for international terrorists because, in the words of Lochrane Gary (2001, p. 1), "it's incredibly easy, requires little sophistication, is inexpensive, and requires little risk of exposure for the terrorists."

The United States as the sole world military superpower would quickly overpower any adversary in an armed confrontation. Consequently belligerents with a grievance to settle through violence have little choice but to wage so-called asymmetric warfare. Such warfare uses tools of terrorism such as biological, chemical, and other nonconventional weapons available to even the poorest of individuals or nations but effective in inflicting damage. In addition to crop and livestock diseases and pests, weapons could include shipping containers for food, or crop-duster airplanes. The latter could originate in rural areas but might be employed to spread pathogens over heavily populated urban areas for maximum impact.

In addition to the above reasons for choosing agroterrorism, terrorists also might be attracted by the damage they could cause. None of the following were caused by agroterrorists, but provide perspective of their potential to inflict harm. They have at their disposal the means to inflict damage at least as great as that for actual outbreaks listed by Kohnen (2000): the Canadian foot-and-mouth disease (FMD) outbreak of 1951-53 costing the country an estimated $2 billion, a swine FMD outbreak in Taiwan in 1996 costing the country up to $7 billion, and mad cow disease costing United Kingdom agriculture $4.2 billion.

Potato blight brought severe famine to Ireland in 1845. Brown-spot disease in rice was partly responsible for the great Bengal famine of 1942-43 that brought death to two million people. The corn leaf blight of 1970 cost the United States an estimated $1 billion. An attack of avian influenza of foreign origin cost Pennsylvania $86 million. In contrast, some 639 acts of terror by the Earth Liberation Front and the Animal Liberation Front since 1996 caused "only" $40 million in property damage according to Federal Bureau of Investigation estimates presented at the U.S. House of Representatives Resources Subcommittee hearings on ecoterrorism held in February 2002.

If agroterrorism is cheap, easy, and so damaging, why isn't it more widely used? Why have there been so few acts of international agroterrorism by al-Qaeda or like groups? One answer is that achieving political change through terrorism requires more than destruction of food and property in a rich country; it requires a shocking loss of lives to motivate a change in public policy.

Chemical, biological, and nuclear tools of asymmetric warfare are most effective in cities because that is where people are concentrated. Agriculture is highly dispersed, creating problems for terrorists who need to contaminate lots of food in lots of places. Contaminating city water supplies is easier. As outbreaks of food poisoning attest, information about bad food spreads rapidly. Also, as the foregoing examples attest, despite massive economic losses from FMD and other pathogens, few lives were lost in affluent countries that have numerous options for obtaining safe food. Thus, agroterrorism is much more attractive to animal rightists and other radical groups in agriculture who seek to destroy property than to al-Qaeda or other international terrorists who seek to destroy lives.

**Response to Agroterrorism**

Combating agroterrorism entails the usual appeals for alertness by citizens and officials, control of borders and resident aliens, better coordination among law enforcement and intelligence gathering agencies, and an end to senseless policies that invite terrorism. Farmers need to be observant of and report suspicious activity of potential terrorists. Greater care to avoid contamination of crops and livestock by visitors is good advice whether the visitor to the farm has good or bad intent. Farmers can help by quarantining animals brought from elsewhere before they join the herd.

It is notable that just one month before the World Trade Center destruction, the Agroterrorism Prevention Act of 2001 (H.R. 2795) was introduced in the U.S. House of Representatives. The bill increased criminal penalties and civil remedies for acts of terrorism against plant and animal enterprises. The bill also established a clearinghouse of information on acts of agroterrorism. Additional measures were taken in the Public Health Security and Bioterrorism Preparedness Act of 2002 passed by Congress and signed by President George W. Bush.

The latter act broadened responsibilities of two agencies of the U.S. Department of Agriculture that play especially important roles in thwarting food terrorism. One is the Food Safety and Inspection Service (FSIS) and the other is the Animal and Plant Health Inspection Service (APHIS).

FSIS inspects meat, poultry, and eggs products sold in interstate commerce. (State agencies inspect animal products produced and sold within-state.) FSIS also reinspects imported animal products to ensure that they meet U.S. food safety standards. The agency tests for microbial, chemical, and other types of contamination.

Working with the Centers for Disease Control and Prevention (CDC), FSIS conducts investigations into sources and types of food contamination. It also inspects individual state and foreign inspection services (inspecting exports to the United States) to ensure that these services meet U.S. federal standards. To control costs of inspections, FSIS relies on sampling at critical points in the food production chain. The processing or marketing firm may do the actual inspection. The system is quite effective in controlling inadvertent contamination of foods by pathogens, but is not well equipped to control for deliberate contamination because only a small sample of food is inspected. Nature may contaminate randomly, but terrorist organizations contaminate purposefully to avoid detection and take lives.

The Animal and Plant Health Inspection Service (APHIS) also is responsible for protecting American agriculture from diseases and pests. Its efforts are directed at plants as well as animals. It tries to protect against invasive species that could harm American wild or domesticated species, including people. Timeliness is critical in stopping epidemics. Quick response teams such as used by the CDC provide a model for identifying problems so that remedial efforts can begin quickly.

APHIS and FSIS need lots of help from veterinarians, physicians, entomologists, and plant pathologists. Greater effort at training and accreditation of such experts to work seamlessly with federal and state agencies can cut response time and false starts. A computerized clearinghouse system for reporting animal and crop diseases and pests could cut response times to identify epidemics.

Only about 1 percent of farm produce is inspected for food safety. While more resources are needed for inspections, resources will never be adequate to stop all agroterrorism. Thus, public agencies need to work with food handlers, helping the latter to know what to look for and to do some of the inspection on an informal basis that public agencies cannot accomplish.

Finally, science can help by developing better vaccines and breeding resistance into plants and animals. Science also can improve technologies for food safety and inspection.

## REFERENCES

Gary, L. "Agroterrorism: How Vulnerable Are We and What Can We Do?" *South Florida Beef-Forage Program.* http://www.ifas.ufl.edu/~sfbfp/A11-01.html, November 2001.

Kohnen, A. "Responding to the Threat of Agroterrorism: Specific Recommendations for the U.S. Department of Agriculture." BCSIA Discussion Paper ESDP-2000-04. Cambridge, MA: John F. Kennedy School of Government, Harvard University, October 2000.

Monterey Institute of International Studies. "Agroterrorism: Chronology of CBW Attacks Targeting Crops and Livestock, 1915-2000." http://www.cns.miis.edu/research/cbw/agchron.htm, 2002.

# Index